❖Fear Not to Sow❖ Because of the Birds

*Essays on Country Living
and Natural Farming
from Walnut Acres*

by
Paul Keene

edited by
Dorothy Z. Seymour

The Globe Pequot Press
❖Chester Connecticut❖

In loving memory of Betty,
who for so many radiant years
shared the journey.

Library of Congress Cataloging-in-Publication Data

Keene, Paul.
    Fear not to sow because of the birds.

    Includes index.
      1. Organic farming—Pennsylvania—Penns Creek.
2. Farm life—Pennsylvania—Penns Creek. 3. Walnut Acres
(Penns Creek, Pa.) I. Seymour, Dorothy Z. II. Title.
S605.5.K44   1988   631.5′84′09748   88–30907
ISBN 0–87106–666–1

First Edition/First Printing
Printed and Bound in the United States of America

Contents

Foreword   vii   Preface   ix

Prelude ❖ 1

The Four Seasons at Walnut Acres

Winter ❄ 7

Spring 🌿 43

Summer 🌿 73

Fall 🍁 119

Photo section appears between pages 88 and 89.

Foreword

One day in 1985 I was reading my favorite organic-foods catalog when it occurred to me that the main reason I enjoyed the catalog so much was the prospect of reading the essay it always included. These essays were written by Paul Keene, the farmer who founded the business. (Or was he a philosopher who also happened to be a farmer? Or was he a businessman who just happened to love writing about the land?)

It struck me that these appealing essays deserved a wider audience than the Walnut Acres catalogs could gather. They addressed issues basic to modern life, such as the urge many of us feel to return to a way of living closer to nature. And they reflected the writer's unfailing appreciation for the way of life he chose many years ago. If I and the other catalog readers enjoyed these essays so much (as letters they wrote demonstrated), the general public might, too.

Not until after I had persuaded Paul Keene to collaborate with me on this book did I discover that we had acquaintances in common. For a time, I had lived with my husband on a small organic farm in New York state built up by people who had studied farming at the same school in which Paul had studied. Paul had also been head of the mathematics department at my husband's undergraduate alma mater, although the two men missed meeting by a couple of years.

Although Paul Keene, by achieving fame in his field, has become something of a businessman, this book is not about Walnut Acres, the organic-foods business; it is about Walnut Acres, the country home and the working farm, where more is put into the soil than is taken out of it and where the farmer is thought of as part of an integrated whole. In reading Paul's essays it became clear to me that, through forty years of farming, he has never lost sight of his ideal: to work with nature, not against it.

Finally, this book is about living one's life according to simple, natural, and basic principles, and finding great joy in it. Being an organic farmer is conducive to living such a life. Reading Paul's essays makes us realize that we, too, would like to have lived such a life. And maybe it's not too late!

DOROTHY Z. SEYMOUR

Preface

These essays were written out of a full heart. They tell of a life lived richly, with the soil ever by my side, my companion on the way. Threaded through my years like a golden cord, my contacts with the soil have tied me ever so gently but firmly to itself and its meaning.

We were not always farm people, although my early years were passed in rural Pennsylvania, and my wife Betty was born and reared in rural India. At first we were teachers, both in India, where Betty and I met and were married, and also in the United States, where for several years I was head of the mathematics department at a small New Jersey college.

But my experience in India inspired me to change my life completely. I felt that the people of India possessed something I lacked in my life, and I was impelled to search for the missing quantity. These simple, sentient people "lived, felt dawn, saw sunset glow, loved and were loved." They knew how to accommodate themselves to circumstances. For them, life was more than possessions, and in their poverty lay their richness. After my contacts with Indian life I was no longer content with formal teaching. I had to get closer to the basics of life. I had to become more directly dependent upon the soil.

In 1946, penniless but hopeful, we found a low-cost farm in the central Pennsylvania hills. Close by was the village of Penns Creek. If ever a place had sublimeness, this was it. More than a hundred of God's good acres to have and to hold, on loan from the universe. Black walnut trees were all about. The given name, Walnut Acres, seemed most appropriate.

More than forty years later, Walnut Acres has unexpectedly become more than a wonderful home where we could live close to the soil. It was one of the first organic farms in the country, and our practice of farming successfully without abusing the soil became known to others. Very simply, we have come to believe that healthy soil produces healthy crops. In return these crops, if used wisely, help to promote health and wholeness in human beings. We have never used synthetic or poisonous nonnatural sprays of any kind, nor have we used chemicals for fertilizers. Our whole emphasis has been on treating the soil as the living entity that it is, on feeding it abundantly with natural soil foods, allowing the soil then to feed the plants. We have come to see that the soil is all of one piece with the life it supports. They are strong or weak together.

Over the years, Walnut Acres has developed whole food products that it sells by mail. Several times a year, we mail a catalog to subscribers. To live on the soil is to learn its secrets. One cannot keep these to oneself. The beauty must be shared. And I have felt inspired to write for each catalog an essay on one among a great variety of rural subjects.

This book reproduces some of my periodic essays, selected from those written over the years. The date on each essay tells when it was originally written for the catalog.

With the publication of this book of my essays, my heartfelt thanks go to my unusually perceptive collaborator, Dorothy Z. Seymour. She suggested the book in the first place. Her appealing offer to select the best essays, to do the editing, to plan the format, and to find a publisher could not be resisted. Only now do I realize how much in need of her attention were my untutored habits in syntax and word placement.

I am grateful, too, for the invaluable work of Emma Mattern. For many years she has dealt capably with everything having to do with the essays and their use in our catalog. I am also obliged to Carol Erb for her dedicated work in committing this material to the computer.

Most of all, I acknowledge constant support by my wife, Betty Keene, who was unable to live to see published a book that is in many ways a celebration of what we have been permitted to share with the universe.

Prelude

FEAR NOT TO SOW BECAUSE OF THE BIRDS

This inscription was found on an old tombstone in a country churchyard. We have adopted it as our farm motto. We have tried to sow enough for birds and people, and then to move through our days trustingly.

INSPIRATION BY GANDHI

"How can a young person best serve humanity and his world?" was my question.

"Ah, my friend," came the answer, "when you return to your home in America, you must give away everything you have. Don't own anything. Then you will be free to talk and to act. Doors will open for you." Mohandas K. Gandhi was suggesting a recipe for my future.

Mr. Gandhi, shining eyes peering over his old-fashioned glasses, had not answered my question, but he had lifted it and me to a higher plane. He had stated a principle, set a direction that I have tried ever since, sometimes in faltering ways, to follow. I have certainly not adopted his counsel completely, but it has set a lifetime pattern for me.

Gandhi had been one of the highest-paid lawyers in India. When I knew him, in 1939, he lived in an extremely simple earthen house, with no tables, chairs, beds, furniture, plumbing, cutlery, dishes, or electricity. I recall seeing only one picture inside his house, a head of Christ. His food and clothes were minimal. If anyone gave him anything he simply gave it away; he owned practically nothing.

When I met Gandhi it was holiday time at Woodstock School in the Himalayan foothills, where I was teaching under a two-year contract. The British Empire was tottering under relentless cries for independence from the Indian people. The Congress Party was hacking at the roots of India's subservience. I had hung around the edges of Congress Party meetings, had come to know a number of the leaders, had sat after hours at the shoeless feet of Jawaharlal Nehru as he relaxed of an evening and poured out passionately his thoughts and hopes. He had left no doubt in my mind about the true greatness of Mahatma Gandhi, in his position at the very center of the whole independence movement.

On the day I asked Gandhi my question, we were walking together along the dusty road he took for his daily constitutional. The countryside was flat, almost treeless. Goats and cattle roamed about. I was on leave from my position of teaching mathematics in the States. After years of preparation and teaching, my work there seemed somehow flat and empty. An unreality about it gnawed at my spirit. Had I become too separated from life at the roots?

It was Gandhi—his simple life, his powerful personality, and his philosophy—who inspired me upon return to the States to spend four years studying and learning homesteading and organic food production. Thus prepared, with only a few dollars, some ancient furniture and farm equipment, and a team of horses, the family began buying a lovely farm called Walnut Acres. Since that beginning, things had always come as they were truly needed. A surprised observer, I have been swept along by life as in a miraculous stream.

I have found that answers do not come by concentrating on one's own desires or fancied wants or needs. Somehow, by seeking out the larger framework, as Gandhi did, one rises here and there above the choking limits of self into a freer, fresher atmosphere, to where one simply sees farther, through an expanded, more beautiful landscape.

We have seen the wisdom in Seneca's remark about the two ways of being rich: One is to have much, the other to want little. I have tried to lean toward the latter, to help avoid the ensnarement Wordsworth refers to:

The world is too much with us; late and soon,
Getting and spending, we lay waste our powers;
Little we see in Nature that is ours;
We have given our hearts away, a sordid boon!

I began to trust more in the universe, giving up a good, secure teaching position to move into the unknown world of organic farming and natural, unsullied foods. I have failed miserably at times to live up to my ideals, but generally with a pricking conscience.

As Walnut Acres grew through our attempt to cooperate with nature and lean upon it, we found that thousands of individuals and families everywhere came to lean upon us for a portion of their sustenance. This priceless burden of trust calls us to ever higher standards. To deal justly with the holy earth, with our foods, with the persons who work so hard to grow and prepare them, and with those whose lives depend in part upon us, is much more costly than producing food in the cheapest way. We have tried not to ask too much.

Never in our wildest dreams of forty-five years ago could we have foreseen what the years would bring. In our chosen work we have been greatly upheld by life. Had we known beforehand what lay ahead we would have been scared to death. But nature kindly keeps our destinies from us. One grows but gradually into one's future.

The
Four Seasons

at
Walnut Acres

Winter

❄

Winter on the old Pennsylvania farm has its pleasures for the eye and the mind. In its many demanding phases it causes us to react to its blusterings by drawing forth our best efforts.

COLOR SEPARATIONS ❄ JANUARY 1983

January in Penns Creek. There against the gray mountains stand the village houses, blending into the bleakness, lost in the flat and dreary season. The land seems almost a lifeless moonscape. The sky has drooped down, wrapping everything in its frostiness. This is our view from the old barn. We could be a bit depressed if this were all we saw.

But the scene is deceptive. After thirty-six winters of looking out daily on the same view, we are not so easily misled by its somberness. We are not tricked by having our moods set only by what the eye can take in. In our hearts we can see the greenery that lies just beneath the surface under bark and snow, awaiting the south wind's call.

For winter is really only one sheet of the color separations that make up the year. In our mind's eye we have only to add the other months, with their color and life, much as a photoengraver does. Then we see this season, too, as a part of the whole. Without the wintry contrast, the composite picture of the year would be incomplete. We welcome winter's peculiar delights. What would life be without its tingle and glow? In those individual houses, beneath gray curling woodsmoke, lie fire and warmth and cheer. Life carries on under its wintertime restrictions, with change of pace and direction, perhaps, but burgeoning and full.

ORGANIC FARMING ❄ JANUARY 1960

Nature understands no jesting; she is always true, always serious, always severe; she is always right, and the errors and faults are always those of man. The man incapable of appreciating her she despises, and only to the apt, the pure, and the true does she resign herself and reveal her secrets.

—Goethe

Back in the hills, fifteen miles southwest of Lewisburg, lies the quiet little village of Penns Creek. The lovely stream bearing the same name flows gently by on its way to the Susquehanna River. Jack's Mountain looms over the creek in a kindly way. Bears still visit the outlying fields. Stories and legends whisper among the trees—of William Penn, of Indians and buffalos and wild-and-woolly outlaws. We feel the strengths and the weaknesses of the frontier town that even today has not yet completely found its way into creative, meaningful living on the far side of the frontier. Yet perhaps we can be thankful for this and the thousands of other little Penns Creeks all over our country. They may yet have something of value to offer that our larger towns and cities have passed by.

Less than a mile east of the village, reached only by a country road in middling repair, one finds a farm named Walnut Acres. Signs direct one there. From the outside the farm looks average, raising average crops in the usual way. This farm has few fences, however, for the last cow has long since bellowed her last bellow.

You should see the barn. From a distance it looks like any other old Pennsylvania Dutch barn, big and reddish. As one draws near, however, one hears the hum of motors. Entering, one sees painted interiors. Gone the horse stalls, the cow stanchions, the manure alleys. The gloom, the spiderwebs, the lowing, the munching—the sights and sounds that for generations tugged at the hearts of old and young have at last given way.

This is not an average farm, however. These are not the usual synthetically fertilized local fields. Ever since we moved onto Walnut Acres, peculiar things have happened. No chemical fertilizers, no poisonous sprays have been applied in the fields, for we had studied a different method of farming known as organic farming or natural farming, and we were convinced of its merits.

Briefly, in organic farming farmers give more than they take. (This is the basis for all sound living, isn't it?) We feed the soil, trying to bring it to a high state of natural, well-balanced fertility. We want to be able to sow and then to sit back in calm assurance that a good harvest will follow, without our having to depend on the shot-in-the-arm effect of synthetic fertilizers. What an insult these so-called fertilizers must be to the life of the soil.

And we believe that in good time, as the soil becomes healthy and independent, disease and insect infestation will lessen naturally. We do not say that the fields will be without insects, nor do we anticipate an easy road. This attitude does seem to make sense, doesn't it?

Many people are looking for foods that they know have been raised naturally, without unbalancing synthetics. They want these foods to be as close to the original, natural form as possible. That desire makes sense, too. Why take apart a grain of wheat, remove the elements that everyone knows are the most valuable—the bran, the wheat germ, the middlings—and then use only the unbalanced white flour? Are we smarter than the Creator who gave us the whole grain? Can anyone really give a sound reason for altering the natural product?

And yet every day most people eat highly refined foods, which means foods greatly out of natural balance. And why? Well, it's a habit—these products are cheap, easy, readily available, they keep forever and a day, the label on the container is so persuasive, and on and on. And often, to our detriment, we quicken and cheapen and submit to high-pressure advertising, and we find ourselves at the same time wondering why we degenerate as we do. Oh, some of us live long, yes, but so often it's a dragging, nagging kind of living in imperfect health, a kind of living we would almost rather be without. Mere length of days without decent quality of living can be unspeakably empty and meaningless.

BOYHOOD WINTERS ❄ JANUARY 1973

One instant I sit and write of the Walnut Acres of today. One ump-teenth of an instant later I have slipped backward through a billion bygone seconds to the whooping days of growing up. Was it long ago that simplicity and frugality were the essence of life? How quickly we adapt to the new and forget the old.

There we were, a warm, close family, struggling mightily to keep hunger and winter from tearing us apart. Incredible strength, society's backbone, was built up under the adversity of early days. A straight pin was meant for saving. A safety pin was a minor treasure. Worn, torn shoelaces were forced into further service by being knotted here and there. The accompanying lumps were so arranged as to lie painfully hidden beneath the closure edge, to become an every-step reminder of the pangs of privation.

A little boy's long black stockings, worn with wintertime knickers, gave up first in the foot, of course. How the toes rebelled in soreness when the darning began. When the originals, already darn upon darn, could no longer be rescued by using the old door-knob darner, the desperately ailing foot section had to be cut off. For some strange reason, black replacement feet seemed usually to be unavailable, but white ones were always to be had. Often, after a cold night's sewing by the old coal-burning kitchen range, with the whole squirming family playing games on the floor, or sitting at the table reading, totally renewed stockings would appear next day in the bureau drawers of the frigid bedrooms. A few sags lay here and there. The knees were a bit stretched, perhaps. But the shining new, white feet promised several weeks or more of additional wear. We walked a lot in those days. We had, simply, no other way of getting about, unless we could afford a bicycle, which we could not. The school might be a half mile to two miles from home, depending on one's grade. Many pupils walked farther than that.

Of course, shoes never quite successfully covered the tops of the new white feet, especially if the stockings were not to be left sagging in limp doughnuts about the calf. Even the high-button shoes worn on occasion were lax in this regard, and so one's pride and feelings were sadly tattered. No matter how hard you tried to scrunch the white feet into the dark confines of the shoes, that white ring seemed always to gather around the ankle. Imagine what white-footed black stockings can do to a growing boy. Eventually, of course, the whole structure of trappings that were once hose gave up the ghost. The limp, shapeless, lifeless mass ended up in the rag-bag, to be given one final chance at usefulness. For Halloween dress-ups, perhaps? Or stuffed with paper for scarecrows' legs?

Winter nights were often spent struggling with the complexities of our one venture into industrial capitalism. One of the maturing siblings decided to break out of the iron circle of deprivation by going into home-style sweatshop stocking manufacturing for profit. Brochures describing the possibilities were insistent with promise. The family, treading on such strange ground, rebounded from fear to hope to sheer disbelief. The shiny contraption to be used for manufacturing, having finally arrived, was unpacked and assembled. One clamped the heavy, hand-operated unit firmly to the kitchen table, cardboard separating clamp from wood. With the clacking monster came spools of woolen yarn. We put one of these spools on the machine, threaded it, and away it went.

In theory, each of the circular forest of needles would rise in stately precision at its predestined time. It opened, grasped the thread, which just happened to be there then, and closed. The captured thread was then miraculously intertwined with another thread, and a stitch was born. Clack, clack—stitch, stitch. Gradually, from the bottom of the gaping round maw, a woollen cylinder of varying uniformity would thrust its being, to gasps of incredulity.

In those days all gentlemen must have had legs that were perfect size-eleven cylinders. Let's not get into heel-and-toe affairs; that was something else, where motherly skills were called for. A special attachment at the top—cantankerous, ornery, and stitch-dropping—provided the ribbing effect. From several to many hours of giddy, eyepopping crankings were needed to complete one pair, plus the final more-or-less futile attempts at shaping and pressing. When twelve pairs of stocking clones were ultimately assembled, they would be packed most carefully and sent to "the company"; we paid the postage. In due season, back would come sufficient yarn for another dozen pairs, plus a check for a dollar and fifty cents.

Just at the time when enough pairs had been made to provide the final payment on the machine, the thing went into a rapid, mysterious, utterly irreversible decline, falling apart, its useful life ended. Packed carefully in its sturdy wooden cradle-coffin box, it sat broken on the attic floor for many a year. The box was too good to discard; it might come in handy some day. The needles became absorbing playthings for neighborhood children—sad commentaries on how the mighty fall. We opined endlessly on the frailty of depending on the written word and on the big-city's reaching into the homes of the humble.

Then there was the after-school picking over of the stove ashes for bits of partly burned coal. For every good-sized bag thereof, taking several hours to retrieve, we received one penny.

During World War I, we spent many a winter's eve knitting scarves for the soldier boys. Ours were mostly multicolored patchworks made from odds and ends of frazzled, unraveled sweater bodies that had separated cold from warmth for many a hand-me-down season. Sleeves were long since gone. As youngsters we were far more enthusiastic than able. We often wondered what unfortunate Yankee Doodle Dandy would have doled out to him our meager offering of undetermined quality and warmth. We hoped that effort counted toward the final grade.

One wonders at the difference between life then and now. Certainly fewer youngsters today seem to know the value of money, the virtue of saving. Sometimes restraints are few, or nonexistent. It seems we have so often forgotten how to live within our means; we simply spend more than we have. Immediate gratification of every whim seems to have become humankind's chief end.

Certainly too, in January, fewer youngsters today know the pleasures and joys stemming from reading and games and self-entertainment. Gone is the one heated room, which for so long lent closeness to the family. With the whole house now heated, gobbling up at fantastic rates our nonrenewable resources, we usually go our own ways. The world sneaks into our being on unseen waves. We now run from outside winter dreariness and cold into the very arms of a mesmerizing, pervasive, otherworld genie known as the Great Softener, to sit and be entertained by the hour. This vivid world awaits only the touch of a button to spring to life, come out of its bottle, entwine us in its alluring, effortless, synthetic, struggle-free enticements. Sometimes I think we did more living years ago, had more fun with less, were better equipped to cope with struggle and adversity, knew better how to make ends meet, and developed a deeper sense of responsibility to persons all about us than we seem to do today.

CHILDREN AT HEART ❋ DECEMBER 1956

The first snow fell last night. As we came home from Christmas chorus rehearsal, the swirling whiteness everywhere seemed somehow so appropriate, so good. It has been but a short, chilly step from the last lingering days of a glorious, prolonged fall into this wintry condition. Now suddenly, here in the north, the seasons have changed and a new life has begun.

With what excitement the new season is welcomed by the child at heart. In our modern world, for "modern" people, genuine spontaneous, wholesome, creative excitement is rare. Too often our appetites are unnatural, jaded by a surfeit of the richest, the most extravagant, the most cultivated, and the most sophisticated goods—in food, in entertainment, in social contacts, in many ways.

This change makes us miss something in the essence of life. Oh, we like to read about the beauties of simplicity and the blessings of childlikeness, and occasionally something directs our thoughts back to the naiveté of our forefathers' days. But our more advanced, more mature excitements are much more comfortable. We can enjoy them sitting down.

One wonders how many of us at thirty, forty, fifty, sixty, or more are children at heart. How many of us ever start running out of sheer exuberance? How many skip with joy? Or walk a mile or two or three just for the pleasure of it? Or folk-dance, or swim, or ice-skate, or roller-skate? Or coast down hills on a sled, or ski, or ride horseback? Or fly a kite, or ride a bicycle, or play volleyball or table tennis, just for fun? So often, assorted self-centered fears, aches, bulges, and conditionings place these and similar simple refreshments beneath our dignity.

Simplicity and naiveté are often lacking in our world today. I think Walnut Acres appeals to so many because they read into its work and message something basically simple. People seem to live vicariously through us the type of life they idealize. I wish we could help make more of the values that grow out of this life available to those who seek them. If only each family longing for its own homestead could find a way to get one. The world has plenty of space for more places like Walnut Acres.

Why, all you need are wide fields, luxuriant with growth as the soil comes alive again; purple mountain majesties, with a quiet village nestling just beyond the fields; rich, brown, vital, vibrant soil to be fed, nourished, and cared for, so that in its turn it may truly feed and nourish; a winding, stream-bordered lane, all shaded in summer; trees and undergrowth in natural variety and abundance; wide expanses of sky, a horizon to lift one's eyes to; a lovely stream for boating, swimming, skating; a barn with hay and straw and mangers, and a life all its own; and always fresh, uncontaminated air, pure cold spring water, unpoisoned, naturally grown foods. Simple things like these.

The new season with its new life is here. As soon as school is dismissed, the children's hour takes over. As I sit back and try to picture it from afar, this time comes to mind as a medley of laughter, arms, legs, scarves, jeans, boots, shouts, slamming doors, dogs, cats, sleds, skates, snowballs, snowmen, recriminations, wet feet, reddened hands, chapped ankles, soaked boots, sodden mittens, cluttered floor, supper, and the full tummy.

But the new season brings more than this. At its beginning, the thankful heart sets its house in order. Summer has ended. The plenteous harvest has been gathered in—a harvest of lightheartedness, of less rigorous moments, of outdoor relaxation, of playtime with the children, of work in the good earth, perhaps of rich experiences in unfamiliar settings.

Chambered in its innermost recesses lie these experiences, to be drawn upon as the need arises. They wait to be transmuted by celestial alchemy into thoughts and deeds of understanding and kindness and forgiveness and love. The overflowing of the heart's bounty in gratitude for countless blessings—for the privilege of life in time of crisis, for glorious opportunity, for high adventure, for deep satisfactions—this is the true Thanksgiving.

To those who stay here in the winter, the cold weather speaks a language of its own, really a language of the heart. The season speaks of childlike excitement and anticipation at a moment essentially spiritual. At this time of year many seem to see most clearly God's influence entering into their lives. No wonder we feel excitement. No wonder children and adults have so much in common at Christmas.

For so often at this time the child's eye, in simple faith, looking up, meets ours in a glance of common understanding. We feel an unaccustomed tenderness and softness, and for a moment we become one. Some of that wisdom of childhood that has apparently been submerged in the work of growing up comes back in matured form. Again we see and understand and believe. When faith enters, years depart. Measured by the yardstick of eternity, there is no age.

CANINE CRONIES ❄ JANUARY 1979

On the field path to the cabin I hear Tasha yipping at Meg from opposite ends of the nethermost pipes of the winter-stacked irrigation system. From thirty feet apart they speak to each other in reverberating tones of the bunny sitting secure within one of the pipes, not even worrying. Let dogs appear, and all the neighborhood bunnies make a rabbit-line for the pipe pile, there to sit chewing the cud, so to speak, snug as a bunny in a pipe. It must be a claustrophobic, deafening, cylindrical encounter, but it seems all a part of the day's experience for hopping, nibbling, nose-thumbing, cottontailed creatures.

In the vicinity of Walnut Acres lives a group of generally unrestrained, free-spirited, hedonistic canines who love to visit the neighbors in turn, showering their muddy affections copiously upon them. They come from all sorts of backgrounds, mostly nondescript, and are largely what is spoken of in cattle-breeding circles as "grade" animals, being not exactly "purebred." Others have more definite pedigrees. Taken together, and at their best, they make up a charming reminder of what life might be like if we, along with the dogs, sitting at our spinning wheels, considered the lilies of the field and the sparrows of the air.

Tasha is an off-white Eskimo type. More aloof than some, she sometimes doles out her affections with a miserly paw. She is quite a huntress. The wee beasties shiver in their uneasiness when Tasha gets on their trail, and they aim for the irrigation pipes. Even that pestiferous rodent, the fierce-toothed woodchuck, hesitates to face her in combat. At times Tasha roams far and wide, visiting the neighborhood feed dishes just in case a little something might be left over for her. Any hesitating, persnickety, or off-its-feed dog might just as well forget supper should Tasha appear on the horizon with the setting sun. To those who know her, however, she is sweet and affectionate.

Duncan is the eldest of the lot. A perennially groomed patrician collie, he seldom strays from his home grounds. His age-hoarsened woofs accost all two- and four-footed passersby with distant geniality. Discretion is advisable, however, for any local hounds who figure on taking a short-cut across the neatly trimmed grass of his well-marked home turf; this incursion he does not care for. He can still give good chase, head and tail high, banners unfurled. He is accommodating in the extreme should Meg, the neighbor's lady dog, and her home folk desire puppies above the common cut. His veins pulse with the purest of blood as, at times, he seems to know well.

Meg is a pretty, young purebred collie who loves to play with others, including small human beings. Sometimes her need for affection appears overwhelming. Walking by one's side, she will suddenly roll over directly in one's path, feet in the air, looking slightly silly. If one then steps over or around her and proceeds, she may just lie there, thinking things over for a time, wondering what she did wrong. When joining the gang of five, with no human adults within smell, she has a rare good time. Something unplanned happened several months ago, a more or less natural outcome of the close fellowship among local dogdom. After gradual belly swelling she suddenly came forth with thirteen wriggling, mostly black pups, to the utter delight of the small fry. The experience much matured Meg, and she will doubtless know better next time.

Which is probably more than may be said of Ralph, the only dog in the region who can pronounce his own name, and often does, in stentorian blasts. He is a gentle giant of black Labrador extraction. Open, frank, fun-loving, uninhibited, always good-natured, one may think of him only as a sweet dog. Never pushy, he will not force his mighty way in if a smaller dog is eating. Rather, he stands back and waits. He'll watch Tasha clean up his plate with neither growl nor whimper, in every circumstance the gentledog! One must envy him his unruffled evenness of temper. He is the spirit of life personified. Dogified?

His constant companion and house mate is one of those small longhaired terrier types who are everywhere all the time, with springs for hind legs. Fluffy aims her nose at you, and thus may be presumed to be looking in the same direction, but you'd never know it otherwise, with that brown mop falling from the crown of her little head. She too appreciates attention and bounces boisterously about one's legs to get it. Hold her front paws and she will totter along, bursting with pleasure, overjoyed, enraptured. She must take four steps to each one of Ralph's. When they run together, she becomes a flying little hairy body with just a blur of legs. What drive! If one pays attention to big Ralph, she'll dance about his head, nipping his ears half-playfully, wanting a small share of all that love.

Rarely do these animals utter a cross word. Rarely do they snap or growl or berate one another. Life is too rich to spoil with imagined slights, with snide thoughts and snider remarks, with attempts to outdo each other, with petty jealousies, with suspicions of malcontent, with fears for the future. They accept one another as they are. Trust, love, the laughing spirit, the sheer joy of being alive—these make them what they are. Other dogs are active in the vicinity, but somehow these five rise above the rest in breadth of understanding and acceptance. With all their ragamuffinism, they have managed somehow to tap the secrets of the universe in unusual depth, leaving us mere mortals so often on the outside looking in. How long, oh Lord, how long?

THE QUIET BIRDS ❋ JANUARY 1961

The quiet birds are here—cardinals and titmice and nuthatches and chickadees and all the rest. One gradually becomes aware of them: a flash of red in the old pear tree, a black-capped head appearing for a moment at the bottom of the lowest window pane, beady black eyes perhaps beseeching or thanking. In a strange, inexplicable way a large green-and-white box among the trees spells "food" to them.

How good of those queer, enormous, wingless creatures indoors to spend months tilling the soil, harvesting the beautiful grain, and then sorting it just to put it out where it can be had when needed. People must be almost as nice as birds.

With crops full to bursting of concentrated sun-energy, how warmly the birds will sleep tonight in their feathers. Of what will they dream?

JINGLE, BELLS! ❋ OCTOBER 1983

One evening, in the heat of a muggy August, I had sunk into a pre-bed reverie, reviewing the day's happenings. Night had almost entirely slipped into place. With a start, I realized that angel bells were tinkling beauteously at what seemed a great distance. As I struggled into actuality, the reverie departed forthwith. Next came tinkling children's voices from the direction of the kitchen door, and suddenly I knew whence the melody came.

There on the porch stood a mother and several small children, just in from Princeton, wondering if they could tent in our woods. Instead of knocking, they had shaken the long string of sleigh bells hanging by the door. As they camped that night in a lovely starlit spot in our quiet countryside, my mind followed the train of thoughts evoked by that far-off fairy sound.

In the time of my childhood, sleighs were still a fairly common sight. In the early days we saw very few automobiles and no enormous, brutish snowplows. Snow packed down gradually, as snow should, and remained on the roads until such time as nature decided to remove it. Roads remained far prettier for a much longer time.

Those were the days of jingle bells. How can one describe that otherworldly sound of jingling bells on trotting horses? The bells were of varied sizes, with assorted tones, ringing out a sweet symphony of joy on the silent air. Roads were snow-hushed, horses (lean and lank) made scarcely a sound except for a muffled, far-off "clop-clop."

Pedestrians used the roads for walking, for there the snow was generally packed tight by horse hooves and sleigh runners. People spread all over the road in gay, jolly abandon, moving to the roadsides only when the tinkling bells suggested they should. Children tied their sleds behind sleighs for marvelous free rides. The beauty is inexpressible.

We still possess a conveyance that was once a fancy two-seater sleigh. It is brought out only at Christmastime, but no horses are there to pull it, and anyway snowplows clear the roads in far too short a jiffy. Life today is often too much of being about one's business of getting and spending, preparing to live rather than truly living.

Years ago we used a large farm sleigh to haul wood from the woodlot. We did not dream of today's fuming, roaring four-wheel-drive pickup. Things were slower, quieter, sweeter-smelling, and simpler in those days.

Near our house in a small town was a mysterious, fascinating wooded hillside, covered with huge old trees. Indians, pirates, and Martians hid just out of sight everywhere. A steep hill lent enchantment, for the snow between the trees could be packed by the tread of many boys' feet into a dandy slide for our sleds. Near the bottom of this sharp slope was a split-rail fence marking the edge of a road.

The daredevils among us found that on arriving swiftly by sled at the edge of the fence, by putting our heads sideways and lying on one ear we could just get under the lower rail at one spot. No helmets in those days. We were heavy enough to get a tremendous start at the top. If we neither flinched nor panicked, we could hope both to get under the fence with upper ear intact and to fly across the road, ending up under the porch of the house opposite the slope. The cries of approbation on running the course bravely and successfully warmed the heart's cockles all through the ensuing night.

No guards were needed, for automobiles in those days were so few. Until, that is, the day when my turn came to challenge the world. All the rituals abided by, this was to be the winner of all winners. The start was ferocious, the track perfect, the fence cleared. And then—oh horrors!—a puffing shadow loomed from the right. Nothing to do but to challenge it head-on—that is, my head on its rear-axle nut. Seven stitches and a two-day coma later, I awakened to a swathed and pounding head and the knowledge that I had taken my last ride down that lovely hill. But it was worth every pang, to be king of the hill for that year.

DEEPEST WINTER ❄ JANUARY 1970

The little birds seem to have no feet today. Their incredibly thin, stiff, bare-wire legs and toes are swallowed up in downy feathers. The warmth of the sun comes to them in the seed they eat, extracted by digestion and pumped by the heart to the feathers caressing the lean shanks.

Legless bodies somehow attached to tree or feeder, there they sit: pert, nervous, bright-eyed blobs of food-seeking life. Think of the provision that permits them ever to carry with them the airy coat of insulated outerwear, its effectiveness regulated at will as the occasion suggests. To be able to repel subzero blasts by simply angling out the downy shafts, creating a thin film of life-preserving warmth —so diaphanous a barrier standing between the cold of death and the glow of another springtime—the sheer beauty of the grand design fills one to the overflowing that is inner quietude.

It has been cruelly cold for several days now—at times leaden, at a times blustery bright. Deep snow lies all about, making life incredibly difficult for the wild things. The feeders around the farm are crowded constantly with many species of birds, up to a dozen at a time. A bit of squabbling goes with it all; the bluejays and the grosbeaks have a demanding way about them. Never before have we had so many varieties, so many birds.

Two nimble red squirrels steal some of the food, weaving in and out of the fluttering multitude on the ground, seeking out first this feeder, then that. We surprise them sometimes by putting nuts about. The feeder at the top of its slender steel post six feet above the ground, guaranteed to be squirrelproof, seems to be their favorite. When we first saw a squirrel in this one we thought he must have leaped from a nearby tree. We soon learned the truth: He scrambled straight up and down the steel rod as if it were a tree branch. Then up over the edge of the tray he slipped.

One feeder is cleverly constructed with a wooden perch as a part of the apparatus. Whenever this perch is jiggled, seed drops from an overhead reservoir into a pan below. The birds are supposed to grasp both the perch and the idea and to work for their food. Over the years we never knew any bird to accomplish this trick. Until this year.

From the kitchen one day we heard a new sound. It didn't fit anywhere among the assortment of household noises that whir on and off or on and on. Suddenly our searching eyes were caught by the capers of a female sparrow who had learned the secret. With great zest and seeming glee she alighted on the perch, grasping it firmly. She had devised a timber-shivering routine of agitated wing-flappings that shook the little feeder from stem to stern. She set up vibrations, much like those of a madly driven sewing machine. The seed poured into the tray. Down she went among her waiting companions to share in the feastings. Then back again to her own special chore, as the other little birds stood about in seeming mute admiration and self-deprecation, waiting wide-eyed. She emptied the feeder.

Some of us here have been feeding a covey of forty quail, a small flock of turkeys, and hundreds of pheasants, mostly females. These poor creatures would starve to death without us. In the deep snow they can find practically nothing. Rabbits have chewed the bark from young tree shoots sticking above the snow; many of these trees will die as a result. We put out waste carrots, hopefully. We are told that deer are being fed in the nearby mountains. These days are not an easy time for any living thing.

In a book more than forty years old, by Henry Beston, a wise naturalist, we came upon this statement: "Nature is a part of our humanity, and without some awareness and experience of that divine mystery man ceases to be man. When the Pleiades and the wind in the grass are no longer a part of the human spirit, a part of our very flesh and bone, man becomes, as it were, a cosmic outlaw, having neither the completeness and the integrity of the animal nor the birthright of true humanity. Man can either be less than or more than man, and both are monsters, the last the more dread."

We cannot help feeling that the 1960s was the decade in which we in the so-called developed world reached a turning point in our lives. In one sense we attained our noblest heights, but we also felt a chill emptiness overtaking us. We learned that we were fast becoming "the more dread" monsters, in our thinking that we were more than human. The end of our monstrosity is not yet, for we are prone ever to hug our toys to our chest. Yet we cannot go back. Nature and humanity together will once more win out. Always we seem to be drawing toward a norm, no matter how far we stray over the short span. We seem always to find a Guide to save us from our monstrosities. Perhaps we shall even grow out of the dementia of fighting and killing our fellows.

The 1960s saw the Kennedy brothers and Martin Luther King slain. These years saw too the Vietnam tragedy. At the same time, the work of Rachel Carson and her *Silent Spring* began to suggest a growing awareness, a built-in rectifying spirit of common sense that seems to be ever a part of life.

You and we also have been in our ways a part of this development. We've tried to be faithful to our vision over the years, though often suffering ridicule and misunderstanding. In our deep feeling for the health and wholeness and purity of our soil and our food, we have turned to nature for guidance and inspiration. Our turning back to simpler themes has been a forward movement.

Through it all we've felt ourselves a part of a tremendous ground swell that will surely sway the future. You and we together must continue to encourage one another, for we are all children of one Nature and of one God. This urge, and nothing less, is the fulfillment of our purpose here.

TINKER DEPARTS ❄ JANUARY 1968

Early last fall our old faithful mongrel hound Wendy lay down in her favorite field and rose no more, slipping from earth-sleep into whatever friendly dogs retire to when they no longer awaken here. Her little old gray, disheveled dust-mop friend of many years and countless jaunts, Tinker, diminished, grew older and sadder and more withdrawn over the months of her long absence. Last night Tinker left on an inspection tour of the farm, over secret doggy paths made familiar by thousands of treks, to make sure, before retiring to his mat, that the borders of his domain were secure against all comers. Today, alas and alack, he lies stiff and cold. The two grown pups sniff their farewells. Soon we shall place Tinker's furry remains to rest alongside Ralph, the duck, on the bank of the pond. Ralph was exceedingly fond of Tinker in life, a strange case of unrequited, unreciprocated love.

From puppyhood Tinker was ragged, not overly kempt. He loved manure piles and briar patches. We could never teach him that cleanliness is next to godliness. We always suspected he just didn't care. What he missed in brushings and sweet scenting was more than made up for by endless days of freedom and fun, of roaming and chasing. He never missed a chance for a romp.

Whenever a loved pet dies, a little thorn of pain pierces back through the years. With his leaving a bit of life leaves, a small door closes, and we feel both impoverished and enriched. Perhaps once again Tinker and Wendy are running together, through the rich Elysian fields of the Dog Star.

HOW WE STARTED ❋ OCTOBER 1984

We didn't set out to become farmers. At one time I was head of the department of mathematics in a small college in New Jersey. On a leave of absence from this work I spent almost two years in India, teaching in a mission school, traveling, studying, inquiring everywhere. Three of the fields of inquiry turned out to be agriculture, rural living, and nutrition. I spent some time in a rural school run by Mahatma Gandhi and his disciples. At the same time two British civil servants, Sir Albert Howard and Sir Robert McCarrison, were in India. I never met them, but their work and their writings helped me change direction. Betty Morgan Keene was born, reared, and educated in India, and we were married there.

Upon return to this country, I taught one more year, but mathematics was rapidly losing ground to thoughts of fields and streams, of proper foods and rural living. Opportunity then came along for us to spend almost two years as assistant directors in study and teaching at the School of Living near Suffern, New York. This school, founded by economist Ralph Borsodi, taught a kind of decentralized, self-sufficient, back-to-the-soil, do-it-yourself homesteading. Its ideas laid the groundwork for much of current thinking in this field, and it moved us to the next step: studying natural farming at a full-time farm school.

At this time Kimberton Farms School, near Philadelphia, was the only organic farm school in the country. In fact, it lasted as a farm school only a few years, long enough to give us two wonderful years of work and study under Dr. Ehrenfried Pfeiffer, who had come from Switzerland to direct the school. Dr. Pfeiffer was one of the foremost authorities on natural farming, and the years under him were filled with wonder and revelation. He helped bring all life together for us into a cohering pattern. During our years there we first met J. I. Rodale, who attended a short lecture course and spoke of starting a periodical to be known as *Organic Gardening*. He asked if I would be interested in helping with this journal, but I said I wanted to spend my future years directly on the soil.

After farm school we lived more than a year on a rented farm in

Easton, Pennsylvania. After some tragic losses there from heavy hail- and rainstorms, we decided in 1945 to try to locate a farm of our own where land was relatively inexpensive. This search led us finally to Penns Creek, in a lovely valley by a beautiful mountain stream, and to a farm just at the foot of Jack's Mountain in an ancient sea bed. Indians once lived and roamed thereabout. A sign at the end of the lane, when we had our first view of the farm, proclaimed LONG'S—WALNUT ACRES—PIGS FOR SALE. The Longs soon left, as did the pigs, but the black walnut trees, often squirrel-planted, grew and still grow everywhere.

And so the dream ended (or began?) happily with the move to Walnut Acres early in 1946. We had to live very simply at first, without plumbing, in a solid old house heated only by wood. For several years Mollie and Prince were our only source of power for farm work; they were the horses we had brought along. Sometimes the going was really hard, even for a young family, and it took more endurance and faith than we thought we could muster. Our worries were multiple. We were practically destitute and had to make a living for a growing family. We wanted to raise our family in the country, far, we thought, "from the madding crowd's ignoble strife." We wanted to make our full contribution to the community through church and school. And we wanted both to learn and to teach all we could about nutrition and natural farming.

Fortunately, we had more on the bright side than on the dark. In my mind I can see now the baby sleeping at the edge of the cornfield as we hand-husked the quarter-mile rows of golden grain on a heavenly autumn day. One day of that made up for all the anguish a whole month could amass.

OUR LIVING SOIL   ❄   JANUARY 1966

The earth has scarcely fallen into sound sleep when we dream again of its waking. Was it yesterday we found turnips frozen in the field, from too little attention too late? Is it tomorrow that calls for seed? How much of an earth year is spent in planning-preparing-planting, in tending-harvesting-laying by. The welcome rest within the cycle sometimes seems so short.

Now more than ever of the earth's surface must be dressed here this year. Penns Creek, a lively mountain stream, leaves the mountains not a half mile from our farm. It flows along one edge of the farm, where fields are reasonably flat, fairly large, and not too stony. Farther back, away from the creek, lie steep-sided, bald-topped, tree-rimmed little hills. Here some of the odd-shaped fields are rough, stern, almost dour. Overall, we do not have the richest, deepest soil on the face of this planet.

But our soil is alive! Years ago a bulldozing man, strange to us and our "peculiar" natural-farming practices, prepared terraces and waterways for us. On his own initiative he came and asked what we did to our soil. He had never seen anything like it. When he pushed it into a pile on one side, it ran down the other. All this richness in a field that, when first we came here, knew no end of unbreakable lumps of earth as large as one's head. He said it seemed almost alive.

If only he knew. This same field was starved in those early, lumpy years. More had been taken out than had been returned to the soil. All we did was reverse the treatment. We began to feed the soil until it gasped for breath. Fortunately, when soil breathes here it breathes fresh, undefiled mountain air. Oh polluting factories, combustion engines, tightly packed chimneys—spewers forth of lung-searing, earth-befouling byproducts of humanity's search for wealth and comfort—stay your unrelenting push into our paradise. Let purity, simplicity, and freshness abound, until the sun rises on the day of modern humanity's awakening and humbling.

In an earlier day the fields on the hills were home to many a stalwart family. Needs then were few, wants still fewer. A self-sufficient cycle held days, soil, animals, and people together in hardship, goodness, and joy. In places, now, only bare, crumbling stone foundations remain. One can build upon them again in the imagination, peopling them with the insubstantial dreams and hopes of a departed age. One wonders if the strengths of the pioneering way must vanish with the weaknesses.

Today these same hilly fields yield for us once more rich crops of grain, surpassing in a good season the crops from the lower fields. And the descendants of the frontier now work in factories, dependent but secure. How little we really know of soil and people, of how and why they function, of their relative poverties and wealths.

ORGANIC FOODS ❄ OCTOBER 1984

"Nature provides no written text on her laws. She only smiles or frowns faintly on her subjects, and whispers softly in approval or disapproval of their conduct. Her disciplines seem very mild, even to the most careful observer, but in the long run, continued obedience to her laws leads slowly to great abundance, and continued violation of her laws ends in desolation."

—Author unknown

Determining how much organic matter and ground rock material we have added to our soil over the years to build its present great fertility and natural balance would be an exercise in higher mathematics. We are happy that we did so transform it instead of depending on chemical fertilizers. Each year we use only 60 to 70 percent of our cultivatable land to produce take-off crops. The other fields, seeded to clovers, alfalfa, and deep-rooted grasses, are allowed to grow up lushly. This growth (tons and tons of it) is shredded at least twice during the growing season and left on the soil in a large-scale sheet-composting program. Every year, too, we use on the land manures, phosphate rock, dolomitic limestone, or mixtures of natural substances. No hay or straw are removed (except on the livestock farm, to which the resulting manure is returned), so that we end up putting back more than we take out.

The soil is an endless source of materials, although we add the ground-rock material to supplement these. Using natural farming methods means that minerals are etched from the subsoil by carbonic acid and other acids from decomposing organic matter. How sensible it seems to use Nature's wholesome, balanced methods to renew the soil continuously. How logical to make available large amounts of many natural elements, letting the plants make their own choices. We feel this system tends toward the highest quality, if not always the largest quantity. Our farm is an oasis of a sort, a small stone-beset patch of earth's surface where poisons have never been used and where the soil has been treated with love and respect all these years.

It used to be said that the best manure was the farmer's shadow. There we are again, manuring. Because we have never used chemicals for weed control, we have weeds. In some years they grow beautifully. It takes many hours of handwork in the carrot and beet fields, or machine cultivation in other fields, to keep the weeds under some semblance of control. Young Amish families, children and all, sometimes help us in this project. They come from neighborhood farms, cheering us all with their willingness and their blithe spirits. Of course, this is a costly project, but our program leaves no other way of controlling the weeds. We suggest that you make diligent inquiries into the way in which your carrots are grown. See how many you can find anywhere that are not at least sprayed, when tiny, with Stoddard's solvent or with a naphtha compound to control the weeds.

We used to worry about insects, wondering if natural farming methods would have any effect on them. That was years ago. We think we have learned a few things now, and I am happy to state that we are rarely troubled by them in our field crops. We may see a few here and there on the beans or cucumbers, but generally we do not have to do anything about them. Natural predators get full chance to do their work. Once in a while, in some years, we may be bothered by cabbage worms. We may even use rye flour or rotenone dust to do them in. Rye flour wraps them about in a glutinous mass; rotenone kills them. Rotenone is made from the roots of plants that grow in South America. It is a natural substance that deteriorates in a few days in the air, the rain, and the soil, becoming completely harmless even to cabbage worms.

We carry our principles over into raising livestock and poultry, too. Under commercial methods a number of things are done that we cannot sanction. There, as we find so often, the emphasis is on quick, cheap growth—to get the heaviest animal or chicken in the shortest possible time at the lowest cost. This expediency may lead to such artificiality, weakness, and susceptibility to disease that all manner of props must then be devised to keep the animal standing up and functioning.

We do not of course decry genuine advances in medical knowledge. But when antibiotics, synthetic female hormones, a broad menu of vaccines and medicines must be depended upon as necessary parts of a commercial cattle or poultry venture, it is, we feel, carrying things too far.

From our standpoint, another hurtful thing is the attempt to raise creatures away from contact with the soil. If an animal or a chicken has free range on a broad, clean, sunwashed, stream-bordered grass pasture—which is, after all, its heritage—then the product must be different from today's hothouse creatures. Commercial chickens rarely get to savor earth's delights. They are born, raised, and die in completely controlled chicken factories.

At present we are raising Holstein cattle. We prefer these to the higher-fat Angus or Hereford animals, whose flesh is said to be marbled with fat. Some think of this marbling as the richness that destroys, for excess fat must somehow be allied with weakness and disease. Sometimes after being moved by truck in a wet spring, an occasional animal may come down with shipping fever, a kind of distemper or influenza. It is standard practice in the industry to inject for shipping fever any animal that has been shipped or hauled. On rare occasions we have had to resort to such an injection; then the animal still has a year or longer to live and grow with us. In this time, we believe the blood will have been cleansed, should it have been tainted. We use no diethylstilbesterol, no hormones, no growth stimulants.

We have the young animals out on pasture during as many months of the year as possible. They eat and bask in the green and the gold. We may chop feed for them if the pastures cannot keep up with their needs, or supplement their diet with good hay.

When the grass has retired for the winter, and grain-and-hay feeding begins, for the most part we use our own organically grown grains, to which we add items like sea kelp, bran, and brewer's yeast. We generally grind and mix our feeds right here in our own equipment. We use all our own legume and grass hay, which is, of course, neither sprayed nor treated. Drinking water comes from an unpolluted small stream, or from our own incredible spring. We simply do not know a better or more natural (or more costly) way of raising beef.

Our chickens are raised similarly. For the most part they are fed organically raised grains, with such supplements as wheat bran, brewer's yeast, and ground limestone, plus foods like soybean meal, alfalfa meal, and meat scraps. They too are permitted to get out to range on the earth, among the grasses, in the sunshine, for as long as possible during the year. We start with chicks that are closer than some to the older, perhaps sturdier breeds of yesteryear.

Chicks are often brought low by coccidiosis, a kind of diarrhea that may attack them fatally. We generally mix the prescribed amounts of a medicine with the chick feed for the first few weeks to keep the infection within bounds should it occur. Later the problem disappears, and the medicine too.

Chickens for meat have free run of the fields after the first few weeks. Chickens for eggs have free run, too, until they are ready to begin laying. Then into the large house they go, all uncaged, to have the advantage of the outside run, plus the company of the strutting, seed-sharing, liberated males.

But all these practices took a long time to evolve and to make viable. People began to hear of us when Clementine Paddleford, of the old New York *Herald Tribune,* learned somehow that we were making apple butter in the old-fashioned way over an open fire, from unsprayed apples that were worm free. She wrote about this practice in her food column. Soon people were writing to us and coming to see us for our very first mail-order product. We called it Apple Essence, and made it of not only unsprayed apples plus spices but also delectable wood-smoke odors and gorgeous fall-day colors. What a way to start a life on the soil.

Next came easily shipped products: potatoes, carrots, beets, eggs, chickens. Then we secured a small hand-powered steel-burred mill for grinding grains into coarse flours and cereals. With that little monster to manipulate, no need to lift weights to develop muscles.

In 1949 we purchased our first stone mill and began grinding in earnest. Now we have a number of assorted mills to make all kinds of fresh, whole flours and cereals.

As we grew we moved our milling operations from brooder house to hog pen to the big old barn—all renovated, of course. In 1958 we finished converting this barn into a modern custom-grinding mill and store. In 1964 and 1965 we added an entirely new wing, to house the new mill with its refrigerated storage bins, a complete cannery, food-processing rooms, vegetable rooms, a freezer room, an office, and a lovely new store and lounge for those coming here to see and to shop. In 1972 we built large new storage, kitchen, and store facilities. The fascination never ends.

On our 360 acres of cultivated farmland we raise as many organic grains and vegetables as we can. We have never used any chemicals or poisonous sprays. The soil is improving steadily, and it is a delight to our hearts. The improvement itself is wonderful, but more pleasing still is the knowledge that our ideas and ideals are sound in the main and that the same laws run through all of life. We generally take off only two crops in three years or three in four years. We do not plant the same crop two years in succession in any field. We work into the topsoil tons of green manure, straw, and organic matter. And we feed the soil with organic supplements and manure. In response, it has really come alive.

Our organic grains are now kept in our refrigerated bins—the first and perhaps only ones in a mill anywhere, as far as we know. One rarely sees an insect. If you think this purity is not unusual, then inquire diligently at any flour mill how they control insects. And what chemicals they use, and how often.

We grind flour fresh every day; it does not stand around for weeks, months, or years. The flours and cereals are 100 percent whole, entire, complete. From start to finish, everything is as wholesome as it can possibly be made.

After we started selling grain products by mail, people asked for other natural foods that we could not raise or prepare here. And so to the core of our own grain and vegetable products we added natural foods that we purchase elsewhere. We pick and choose very carefully. We always find out as much as we can about our products. We find completely organic sources whenever we can. We are careful in defining the word *organic,* and we use it only when we feel certain about it. We use symbols to indicate various categories. We want people to know what they are putting into their bodies.

In a sense, we have taken the lead in this field over the past thirty-five years or so. In the years since the first Earth Day, in 1970, progress in this direction has been phenomenal. Truly natural, organic foods have moved far beyond being the fad that some people assumed them to be. They have proven that their growth and use lie in the direction that Nature itself dictates. We begin to see once more that continued violation of Nature's laws could indeed end in desolation.

SQUIRREL CAPER ❅ JANUARY 1981

A fat gray squirrel, hunter-spared, tail-propped, forepaws adangle, stood up among dead leaves to check the log cabin in the woods for inhabitants. Had he been less winter-sleepy, he might have scurried away at the signs of life he detected, but as we watched from the window he merely turned and ambled toward his penthouse bedroom in the nearby old hollow oak.

That reminds me. Years agone, elsewhere, I struggled through cold, fresh, knee-high woodland snow in bright sunlight, noticing a huge squirrel's nest thirty feet above the earth. With a stout club I struck the supporting tree a brace of smart raps, just to see what would happen. I saw: At least a half dozen stupefied, staccato squirrels shot as if by magic from the dried-leaf bower, bounding at crazy angles, legs stretched to the limit, bodies tense in teeth-clenching terror. Through the shimmering ether they arced downward, subsiding gently into the frosty softness beneath their home. A great deal of confused, sheepish thrashing about followed, until all were once more safe in bed, recovering from their unaccustomed daymare.

Being generally respectful of sleepers, I've never since repeated this caper. But I have had many a chuckle when memory has rerun the event in every bright detail. The thought of it now lightens the dreary, drooping day. In my heart is sunshine, with leaping bushy-tailed rodents sailing at the edge of the field of inner vision. I'm content to think that the squirrels, come the next spring, joined me in enjoying the little joke, and even told their grandchildren all about it.

GRAVITY AND US ❄ FEBRUARY 1984

As this is written the wind whistles through its teeth. Blustery turbulence is everywhere, driving the scudding clouds, those little puffs of moisture, the vessels themselves the sails. They expand and coalesce. The sun sickens, its blood thinning, and now tiny, flattened-out, frozen white dewdrops appear from everywhere and nowhere. The symmetrical flakes swirl about one another, wildly. They sail across the fields joyously, competing to see which can longest resist gravity's seductive tug.

Ultimately, as ever, gravity, the great magnet, wins the contest. It enters into the fun, playing the game so as to keep the thrusting, driving flakes from losing face too badly. The force may even loosen its tether a bit here and there to give them minutes more of fancied freedom flight. For them its bidding is ever so gentle but nonetheless persuasive. Eventually Earth wins the game. The flakes flutter a few moments on the ground before giving in to the universal embrace, but eventually they join their companions in their common bed.

What a lifelong struggle every entrant upon Earth's stage faces. Plants lift their heads, strengthen their backs, grow vertically into the air, extending an Earth radius that begins at its very center. It is easier to stand, to resist the inevitable force in that way. Even a broom, straw up, can be balanced on one finger if held perpendicular to the surface of the Earth.

But to reach more of that marvelous sunlight a tree must branch out. In so doing, its limbs are subjected to more of earth's pull than is the perpendicular trunk. The tree therefore sends out its branches in a whorl, to balance its struggle against gravity by distributing the branch weights with roughly even distribution around its central upright. It thus becomes a larger broom.

Further, perhaps for a time to reciprocate for its tendency to pull down, Earth allows—nay, even encourages—a tree to spend its vital days in equilibrium, so as not to be overwhelmed by possible dissolution before its purpose has been served. Thus Earth and tree combine in an anchoring root pattern reaching out in all directions, holding it firmly against every wind that blows.

If a tree is thrown out of balance by losing a large limb in a storm, compensating factors set in to equalize the forces. Perhaps other branches on the damaged side grow larger faster. Or the roots on the appropriate side grow out farther, thicker, or stronger, to encourage equilibrium. The cells at the top of the limb–tree jointure may grow tighter, closer, stronger than those on the bottom, to oppose the pull of gravity, to resist the tearing off of a large limb. When and where needed, the tree muscles grow stronger.

Standing off to the side, one can see in the midst of this maze of opposing forces another wonderful coincidence. Enter the Hand of the Artist. To spread themselves before the life-giving sunlight, branches must reach out. As they do, their ends can least resist gravity's pull. Graceful arches form to delight the human eye. The tree stands midway between life and death. It is kept in wholesome tension by being drawn in several directions at once. Out of this tension grows a beauty that otherwise would not be there. And we human beings are made to know this configuration as beauty when we see it. Earth thus provides us with food for both body and spirit. To a Martian's eyes the whole thing might seem unappealing. To us it is just the opposite. The beauty and the sensing of it were made for each other, for all-around mutual enrichment.

The branches' flexibility that permits them to sway beauteously in the wind, the leaves' ability to flutter at the ends of their rubbery stems, the skirt of cooling shade held close to the dripline like a welcome hole in the summer sunshine's heat, the marvel of the late afternoon winter's light on the bark skins of naked trees, whose leaves have long since been sucked to Earth—these all tell of the beautifying effect of gravity on our lives. What if a tree had the same leaves for its entire lifetime? What if leaves became detached but no gravity were there to pull them to Earth? They would be everywhere all the time, like astronauts lolling head over heels in a capsule in space. Fancy all the leaves of all the millennia still floating about our ears.

Gravity causes there to be beauty everywhere—in the horse's graceful running, in the space shuttle's landing, in our dancing and diving and swimming, in our playing tennis. Practicality lies there, too. We would have to be different kinds of creatures even to exist without the pull of the Earth. It can be both our friend and our enemy. Months of infant struggle are required to tear away from it enough merely to stand and walk about. We hurt when it conquers us and we fall. We find ways to get new hips, new knees, new faces to foil its workings. We spend a lifetime trying to push away its effects. Still, with the years we grow shorter in stature. Stooping and sagging, we show our increasing inability to stay ahead of it.

How Earth speaks to us if we have the wit to listen. Gravity is but Earth's way of holding on to its own. How fortunate we are to be a part of, firmly tied into, Earth's systems. The mind staggers before a vision: billions of umbilical food cords, one to each individual, drawing continuous sustenance from Earth's precious thin skin, each sustained person standing upright, feet sucked fast to the pincushion from which we protrude. In our captivation by gravity lies our freedom. We could not exist otherwise. We and the Earth were made for each other.

A LARGER WORLD ✳ JANUARY 1969

The end of the year just past marked the end of our old selves. We are different persons this year. For out of inconceivable depths of space, at Christmastime, a voice flashed from a space capsule into the ears and minds and hearts of a listening world. With the first shock of incredulity abated, heightened awarenesss crept ever so slowly over us. The doorway of perception had been forced open a crack wider. We can never be the same, for our visions have expanded; we feel more mature in relation to ourselves, to each other, and to our universe. If in a sense we die a thousand deaths each day, and rise better persons, certainly for us in that week shallow comprehension gave way to a new depth of spirit. We were in a sense reborn into a larger world.

How lost we might consider ourselves to be in space. Yet we know now that something is truly unique about this beautiful blue and white marble, coursing through limitless cosmic blackness and void. We know that a universal combination of substance and law makes it possible for us, particular combinations of substance and law and spirit, to live and love on this glorious orb. We know we are all perennial astronauts, encapsuled in our envelope of life-giving, life-preserving atmosphere. We now know the good Earth is our special home, ours to respect and to enjoy. If any marriage was ever made in heaven, certainly this one was. We were made for each other. We will not always violate and ravish our food supply, that umbilical cord tying us ever to our mother, Earth. Nor will we always befoul the capsule within which we move and have our being. And perhaps now because of the sudden growth we have all experienced, the lion and the lamb shall soon lie down together, and we will not study war any more. We are too big now for the little things, the gettings and spendings of our childhood. We must be about more important things, the only things that really matter, the things of the spirit, like love, and brotherhood, and unselfish understanding. We must grow up into readiness for the next big steps as they come along. Through three men God has spoken afresh at Christmastime, making the old story remarkably alive. We can neither forget nor go back.

THE SAP IS RISING ❄ APRIL 1975

For years we've been dreaming of tapping the huge maples that grow along the old lane, overlooking the gurgling little stream. What a place for all seasons: just beautiful. Bob finally sallied forth with some hundred-year-old maple taps, pounding them into holes he had bored, his two little fellows helping. He first hung two-quart metal pails on the tap hooks. Along came a warm sun, drawing the sap from the roots and into the pails. He quickly substituted four-teen-quart buckets for the pails, and overnight they overflowed. Soon milk cans, buckets, every available container was pressed into service to hold the sap.

His first boiling down (driving off twenty-five gallons of water to get every gallon of syrup) in the big old kettle over a roaring fire was great fun. His family treasures the results. Once more Nature's pumps are about their incredible task. The sap is rising.

THE HUNTERS ✳ JANUARY 1957

For a month the small game has been hunted into fencerow cover or holes in the ground. Only now, the hunting over for them, do timorous little inquiring noses dare again to tremble outside in the evening air. For them another year's holocaust has come and gone. Now in peace in winter quarters they will store up for another spring, another family.

Come the hunters, out would rush furiously our individualistic, nondescript dogs, tails in the air. With the first shot they whimpered and cringed back to the shelter of the house. They had not bargained for this encounter with violence; it takes stomach.

On their own the dogs are too slow and doddering to hope to catch anything. All winter long rabbit tracks crisscross the lawn. Here bunnies have played down the lane, run up the bank there by the stream, unafraid and unharried. Dream rabbits are so much easier to chase, and perhaps even more rewarding, for success is more often assured.

Yesterday a lovely doe, wild-eyed with fright, jumped in front of me from a high roadside bank. Today we hear shots up along the mountain. It is not difficult to picture the lumbering bear or the bounding buck, fate met in mid-air, crashing down to rise no more. Somehow life in larger beasts seems itself larger. We may set out the ant trap without qualm, but something about wresting life from larger creatures of the wild causes pain. Something in us responds in admiration to the ability to live close to the earth in field and forest. It hurts to destroy that which we admire.

Strange creatures that we are: to brood over a shot heard on the mountain while thoughts of the morning bacon remain with us. How easy to forget when we do not see or hear or think; how strange that we feel we must exploit life to maintain life.

COME SPRING ❄ APRIL 1962

A hungry hawk circles endlessly above the woods across the next field, flat against a flat blue sky. One can almost feel the intense piercing of those awful eyes. Let little creatures of the world beware. No stirring against the white snow cover will escape that gaze. Spring is coming—but it is not here yet. A few days of warm sunshine, and food and hiding will be easier. Just now hundreds of tracks leading into little round holes in the snow all over the old cornfield tell of fieldmouse activity there. What a lot of friendly visiting goes on.

Of a sudden, in the middle of a field, tiny bird footprints appear from nowhere. A smattering of them around a slim weed stalk tell of the few minute weed seeds standing between life and death. In the fencerows pheasant tracks come and go. The struggle these heavy birds exert against the force that would hold them down is often most beautifully told by a wingtip brush in light snow. One senses the angle of lean, the direction of takeoff, the straining of every muscle in anticipation of that thrust into space.

Rabbits are everywhere, right under the noses of the sleepy old dogs. They are concentrated in the garden area; several dens lie there. With all the clover and grass in the world at their feet, they know where the cabbages are to grow come spring.

Come spring! The heart leaps at the thought. Say what we may, for most of us those words carry the stuff of which hope is made. The farther north we live, the longer the winter, the keener the anticipation. We know what lies between us and spring—the early starts, the false promises, the hopes dashed. We bear these bits of knowledge in common with all of God's outdoor creation. Can leaf buds resist the lengthening day though bitter nights may yet come to give them pause?

Although the force that quickens the sap, bestirring it from its hidden, mysterious reservoirs, may chafe at the sudden freeze, the belated snow, it is not really deterred. It lays its plans, marshals its reserves, and proceeds with utmost faith. We cannot know how or why the great whisper comes, but when it does come, when that something happens for which winter has been but time of preparation, a withdrawing to allow further joyous expansion, no power on Earth can stop fulfillment of those plans. The seed shall sprout, the blade of grass shall rise.

Spring

At Walnut Acres, spring performs a great alteration in the landscape—colors change en masse, the soil gives up its secrets, the wind rises, water floods, flowers burst forth, birds mate, and lambs are born. The purpose in all life suddenly appears to be beauty.

WHISPER OF SPRING ❦ DECEMBER 1960

Here in our little valley, where the world enters over one dirt road, the whisper of spring has come. How appropriate at this late Easter season to have the eternal rising and expanding of spring so obvious all at once, everywhere, in all directions. It is in the air we breathe, the water we drink, the sights, the sounds, the depths of understanding. It is life that rules here and now. Its force is so great at this moment that one can feel it as a determined, unceasing pushing, sweeping all before it, covering, smothering, overtaking, laying claim to all it touches.

Life needs such simple, common objects in which to express itself. Let there be a bare, brown stem; a shriveled, frozen grass blade; the blob of a bulb; an incredibly tiny seed; a gnarled old tree; a bird body, just skin and feathers; a soul within a frame that is but grass. Let there be but these. When the call comes, the mighty swirl is suddenly everywhere.

Oh you chemists, you biologists, let us hold on to our mysteries. Let us see life as life, neither more nor less. Do not tear away all veils and reveal only stark, naked crystals, orbits, and pressures. Leave us with the strength and the assurance of mystery to lean upon rather than the weakness and the confusion of "fact." When mystery is gone, life is gone. Mechanics alone remain. Let God have the upper hand still, and not the human mind.

The wheat has come alive again, and a glorious green it is. Already it moves with every breeze. Below the taller wheat come the grasses, the clover, the alfalfa. As the wheat grows taller, shading the surface of the ground, these smaller and slower plants begin their work of covering the ground. By the time the wheat is harvested they will be secure. And so will the ground.

On looking out over the field one can see not only the green, waving promise. In the mind's eye one can also see the golden brown of the finished loaf, of the promise fulfilled. And so the wheel turns. From brown earth to green wheat-grass to brown loaf. Green and brown—Nature's basic colors.

In a day or two we hope to have the oats planted. Then comes the wait until the myriad light-green oat spears puncture the surface of the field in a million places. Oh, the glory of an early morning dew fixed in a million droplets on these tiny blades. Who cares how or why it happens as long as we can be thankful that it happens? Then corn planting, soybean planting—and before we know it the wheat harvest again. Each step in itself is wondrous, and so are all of them together. We would not have it otherwise.

SADIE ❦ APRIL 1974

Sadie, neighbor Bob's weirdest of English setters, cries for attention each day. She is insatiable. She is also nervous. One little pat, one word of affection, and off she goes into flights of ecstasy. In thirty seconds she has dashed aimlessly up to the woods behind the house, a whole field away, and returned, eyes aglitter, tongue aloll. Then it starts all over again, and continues until one takes refuge in the nearest building. Whereupon Sadie makes the rounds to all the windows, fixing any human being behind them with her beseeching eye, her piercing whine.

All this evasion is preferable to being in the same room with her. Let one sit on a chair and show Sadie the slightest attention, even to the mere catching of her eye—fateful curse! Immediately begins a prolonged battle of wits. She has all night to accomplish her ends. My mind goes back many years to a train ride outside Cairo, in Egypt, when chance placed me beside an obvious pickpocket, who snuggled closer and closer as his hand frisked over my body. Thus Sadie snuggles, insinuating her bony carcass millimeter by millimeter onto one's very being.

She is an instinctive mistress of the feint-and-parry technique, invariably gaining ground in the end, no matter how tumultuous the reaction to her approaches. Once a dog has convinced herself she is starved for affection, no amount of it suffices. That's her trouble. Only when and so long as her heiferlike proportions are finally entangled in and overdrooping one's lap does she show a semblance of inner peace.

Ah well, she is, after all, one of God's creatures. And suddenly now, Sadie also wants so much to be a mother. Heaven protect us. Spring is here.

LIFE KNOWS ITS APRILS   ❦   JUNE 1961

One night two weeks ago a very late, very heavy snow tore out by its weight two of our tall locust trees along the lane. Other branches lay strewn about, showing up weaknesses we never suspected. Wires came down, too. And suddenly we were brought up short. Our self-sufficiency melted, as did the snow the following day. We were isolated: no light, heat, time, water, work, or telephone for almost a day. Candles, fireplace, and dipping in the spring brought back earlier days. Once more we were pioneers. Who cared about time, work, or telephone? The race halted. Time stopped. We could smile and tell the children of the years when we were unspoiled by conveniences. We did not work harder then; we were no more tired at day's end. Sometimes one wonders: At what point does gain become loss?

This memoir is being written in April—fickle month—carrying water on both shoulders, and generally spilling some from each. One day pretending to grim, fierce sternness, the next calling forth buds and leaves and flowers, and birds and insects from their winterings. A past master at cold-war techniques is April.

Time moves on. Snow and slush are followed by mud and dust. These tell about the mixture of feelings that is springtime. Yet with it all we plow and harrow and plant. For seedtime comes and goes swiftly. We have never seen our fields more lush. And nothing can stop them from reaching their determined end. For they are a part of a most beautiful pattern, that of life reaching for its goal.

Life knows its Aprils. It is not disheartened by them. It has lived through so many of them. What will be, shall be. Let circumstances be favorable, let the warm breezes play for ever so short a time; when the call comes, the sap rises. The buds, ever new, know somehow the secrets of other years. Should the cold once more settle down, they shrink within themselves. But rarely do they die. They merely bide their time. They know, and knowing, hold their peace. A few warm hours, and they swell, awaiting their fulfillment, which is beauty's custom. The plan and the purpose abide, outliving interruptions.

Some think that all life is directed toward production of seeds, with the seed the be-all and end-all of the plant or the tree. The species must survive, the race go on. But could it be—oh could it just be—that beauty is the reason for being? Could it be beauty that must go on, with the species, the race, the seed incidental? The plant for the flower, the tree for form and shade, the green fields for surpassing glory? Because of grain and fruit our bodies live. But first, spirit is fed. In our hearts we may decide which is more important, for which the sap really rises. I know where I stand!

FIELD VESPERS ❦ MAY 1967

In this season of frequent showers, with about three million seed peas waiting to be planted, one must take every opportunity to work in the fields. Today my opportunity came. By early evening the earth had dried enough to let me get into the field with tractor and disc harrow.

The square, ten-acre field that I was about to plant lies apart from all buildings, high, dome-shaped, falling away at two corners. To the east, down a beautiful wooded slope, the earth dips sharply, dropping perhaps one hundred fifty feet to a lower level. From the edge one can look far out over the tops of the lofty oaks at the bottom.

To the west a string of tiny village house roofs gleam in the late sun, looking for all the world like the teeth of a smiling Jack's Mountain in repose. Far into the sunset the mountain tapers, bathing soft Teton-humps in liquid sky at the horizon.

As the tractor and I crawl humming around and around the field, with each lap I take in different sights. The wild cherries in the fencerow shower the ground with white petals, so that as I pass by I return to the soil that which so recently came from it, its beauty so soon spent. Maples show the tiniest bit of red life. Sassafras leaf buds are tight little golden balls balanced on twig ends. I tear off several scraps, to chew the bark as we bounce along.

The discs throw the damp earth first one way, then the other. Weak weed-roots have their earth-life torn from them so that they will perish in the next sun. Lumps of soil are broken up. The harrow lurches mightily, screeching wildly as it overtakes and jounces over a large stone. We traverse the clay spots several times—it's lumpier there. Redwing blackbirds flock onto the overturned ground for a last late snack, in ecstasies of hasty pecking.

The east has already been called to the night. Now the west darkens. All the little hills are trimmed with gold. Birds have gone. Suddenly one knows that Venus is there, has been there for a long time. Stars pop out all over. Zip goes the outer coat, down the ear muffs.

I turn on the tractor lights. One in back shines on the harrow; only half the front pair works. A severed wire hangs limp, useless. It's harder to see now. I must always keep the lighted side next to the worked ground, to see where I am going. The turns are confusing.

Darkness moves in from everywhere. I am a little firefly buzzing around in an enormous planetarium. I stop, turn off lights and motor, and just look and feel.

Where there could be emptiness and loneliness, there is fulness and peace. I am a part of it all. I am in it and it is in me. We are made for each other. The smell of the fresh earth rises in reassuring waves. It and I have secrets. The stars take me in until I wonder about other little men on other little one-eyed tractors on other faraway stony fields, harrowing stardust in this glorious universe.

Gradually all over the east a faint flush appears on the black curtain. The moon is a few days past full and will be late tonight, but it comes up, lopsided, as I near the end of my work. It's almost ten now. Temperatures will be at freezing and below tonight. I'm glad for insulated underwear, sweater, coveralls, jacket, ear muffs, and gloves. I go the half-mile home in the moonlight, lights out, singing aloud. This is life. Sleep will be sweet.

OUR STRINGS LEAD BACK 🌿 APRIL 1967

I was out flying my kite the other day, helped by a sturdy young lad. Actually, it was *his* kite, but the wind was strong and gusty and a larger hand was needed. The cord bit into my cold flesh as the tormented bird fought the leash. Little did the kite realize that the pull of the string, which seemed to hold it back from reaching into the blue depths above, was the very thing that kept it sailing. Suddenly the string broke under the strain. Immediately, tensionless, the feeling of life left. No more strain. No more tautness. No more fullness. No more certainty.

The elements were the same. Wind, kite, string, man. But the connection was lost. The once-proud, powerful kite in the next moment was a fluttering, defeated bit of paper, slowly sinking—spiritless, reduced to its elements. Its freedom, its very being, resided in its having been tied to Earth. To be free of its tie was to be lost. There was a sadness about it all that hurt, the sadness of a lesson learned too late. I could not help thinking of that haunting verse that begins, "Make me a captive, Lord, and then I shall be free."

For us living in this wild day, the chips are not yet all down. The second hand of time's clock has barely moved since humanity's supreme enchantment with itself began. Yes, we are cleverly intelligent. We know ten thousand times as much as did our grandfathers. We are wealthy beyond measure. We can pile one costly gadget upon the other in childish bemusement. We have power incomprehensible. You want a nation destroyed? We can do it, now. How impatient we become with the slow, the poor, the weak. Push, push, push. Be important. Get, get, get. Oh, dreadful drive.

Whence this inner emptiness that the pushing and the getting seek in vain to cover? Whence the uncertainties that haunt us to the very bottom of the sea of tranquilizers, stimulants, narcotics, pain-killers, and drugs in which we immerse ourselves? Whence the body imbalances that sour our days and threaten to remove us, willy-nilly, from the scene? Whence the current flight from the soil to concentrations in ever-growing cities?

I don't know all the answers. No, I wouldn't want to go back too far. No, we can't all be farmers. But I believe we can no more overlook our basic connections with the soil and the life it represents than we can escape final return to the elements. I feel more strongly every day that many of our present difficulties arise from attempts to achieve just this escape. We suffer the curse of supposed freedom from restraint.

It seems really to lie with us, whether or not to sever our strings. It is we who determine mainly if we and our children shall flutter aimlessly or strike out boldly. It is we who decide either to grip the certainties or to live with our confusions. All our strings lead back eventually into one healing, holding hand, which seems ever ready to tug us against life's storms into adventurous strength and power.

FLOOD! ❦ MAY 1984

Between its starting point and the north edge of our farm, the limpid, rippling stream called Penns Creek flows for forty tortuous miles along the lowest possible path. Most of the year it gladdens hearts all along its course. Farmers living by its banks find it usually to be a good friend. Because no cities or industries are found anywhere along its course, there is no concern about the possibility of pollution. Its water may be used trustingly for the watering of farm fields.

In hot weather Penns Creek becomes one winding, elongated swimming pool. Old and young alike appreciate its coolness. In its mild embrace our children learned to swim. Many were the picnics and other good times experienced on its banks. Local churches sometimes use it for baptismal services.

Fishermen and women sit pensively by its side, not really caring much about the fish but enjoying immensely the calm and the quiet. In early spring the fish called suckers find their way from the ocean to the place of their birth in the small tributaries winding through the farmlands of our area. Some years ago, high water came while the suckers were running. The stream soon overflowed its banks, spilling right across the main entrance road into Walnut Acres. As the mature fish wriggled and struggled upstream, they swam in hordes in the widened waters right over the road itself. As they did, some enterprising neighbors waded in with bushel baskets in hand. They scooped up hundreds of fish from the surface of the covered roadway. Then home they went, to feed them to their hogs!

Penns Creek arises out of one end of a large cave hidden under a cornfield. It simply wells up, full blown, almost frighteningly, from somewhere underground. Several times when our girls were young we went to see Penns Cave and to take a ride through it by boat. The boat was carried along on the water by that incredible fountain arising from out of the bowels of the Earth.

One enters by going down a long flight of steps into a mysteriously cool, damp, and somewhat forbidding grotto. Often a bit of shivering seems in order, perhaps attributable to more than the sudden coolness. An iron fence at the entrance of the cave, with a locked gate attached, keeps people from falling into the water. Only in loading or unloading the boat is this gate opened. People wait in hushed quiet for the next trip, their thoughts somehow pensive and deep. Even the children seem awed. One thinks of Samuel Coleridge's poem about Kubla Khan in Xanadu:

Where Alph, the sacred river, ran
Through caverns measureless to man
Down to a sunless sea.

On one particular trip to Penns Cave our youngest, Jocelyn, was only two or three years old. We were standing at the fence waiting for the boat. Jocelyn was the baby and looked very cute in her hand-me-down clothes and her grand, new, hard-to-come-by little brown shoes. Having, as they say, a mind of her own, she was determined to stick her foot between the fence bars, just at the edge of the up-welling, vexed waters of the deep. She seemed to want only to touch the very edge of the water with her toe, to prove some point or other.

In the midst of suggesting quietly but firmly that she pull back her foot, I suddenly noticed with distress that her shoe had fallen off and was bobbing about in the swirling backwash just at the edge, getting ready to dash out into the main flow into the cave!

In no time at all I was down on my knees, arm stretched at full length between the bars. As the miscreant shoe danced one final light fantastic before my very eyes, in preparation for its trip into oblivion, my thumb and forefinger managed deliciously, just at the last possible second, to grasp its top. Upon which Jocelyn commented calmly, "Thank you, Daddy."

Not all of us can be so calm in the face of strong and rushing waters. Have you ever been threatened by a flood? It's a challenging happening, to say the least. I must tell you about our experience of the past few months with Penns Creek, which collects water from all along its course through Central Pennsylvania's hills and mountains. By the time it gets here it is quite substantial. Let the rain come down heavily for a spell, especially when the ground is still frozen so that the water cannot sink in, or let a warm spell in winter melt the snow cover rapidly, or let a combination of both occur, and that gorgeous stream changes.

For people who live close to its banks it may become suddenly demonic. One can watch civilization float by: refrigerators, tables, oil tanks, mobile homes, or parts thereof. Let someone say "The river's up" and a cloud settles over everyone's horizon. "How about Joe—do you think he'll get it this time?" Chickens and animals are carted or driven from low-lying farms. Hatches are battened. People move out. That creek can come up faster than anyone can imagine.

The stream that may regularly be a hundred feet wide and two feet deep can in a few hours become in places two thousand feet wide and ten feet deep. Its power becomes absolutely incredible. For years after a splurge like this, one can tell where the water has been by checking the height of the debris caught in bushes and trees, or, if in winter, the height to which crashing ice chunks debarked the trees along the banks.

Well, a couple of months back it happened again. It rose faster than we had ever seen it rise before. In several places the entry road was covered in no time. We had to get out the back way, on higher ground. Debris appeared, rushing madly down the river's surface. People started moving out. One family we know well not only had the usual cellarful of water, but the water came to within two inches of the first (living-room) floor. Family members had moved all the furniture to the second floor, as they have done time and again in the past. Walnut Acres has one barn near the creek, and that evening a number of Walnut Acres men went there to haul the stored potatoes and other products from its ground floor. But the water stopped rising just inches short of the floor, and we were saved.

Yes, we survived, grew closer together, have 1984 to talk about to our grandchildren. The messes are cleaned up, the cellars are less dank, life is back to normal. Perhaps there is even something good about unpredictability—the possibility of sudden unrelenting hard work, of danger and discomfort, of threat and loss. Out of having to cope, and in doing so, one can develop a strength, a depth of understanding, a realization that life does not promise us eternal ease and comfort and release from problems and difficulties. We learn that within us is a power we seldom realize we possess: the power to rise above. We normally use—or want to use—but a tiny fraction of the stamina and sturdiness that are our birthright. More power to us. We're not begging for more floods, but we're also not nervous wrecks over the prospect. Life goes on, people cope, they rise above. That is one of the beauties of the human condition.

IN ITS OWN TIME ❦ JUNE 1968

This is the time of the flowers. Some rare combination of favorable factors has made this year's blossom-burst particularly overwhelming. How truly thankful everything must be to have winter's long sleep over at last. Trees and bushes we planted many years ago are only now beginning fully to show their true glory. Nursery catalogs do a wonderful job of collapsing the years into words and pictures. Young people, do your planting early and well. Make sure conditions are right and good. Plant for yourself and those who come after you. Life's full beauty develops in its own time, and sometimes most deliberately.

FAREWELL TO LASSIE ❧ MARCH 1957

Lassie is gone. Our dear, sweet old Lassie, who in fourteen years saw many a change take place in our lives. She saw the move to the farm—how she hated the trip here in the old car. She saw Daddy Morgan come to us from India and then pass on to the next life— how she loved aged people with their slow, quiet, understanding ways. She saw the children appear on the scene one by one—how knowingly she accepted them, letting them ride or mistreat her with never a complaint. And she always greeted visitors in a most friendly way.

Oh, she was queer all right. Somewhere in her puppy days, before we inherited her from the S.P.C.A., she must have been mistreated. She was afraid. Was it the collie in her that dreaded the summer storm? She became insane with fear when lightning flashed and thunder blasted all about. She would tell us when a storm was approaching long before we would otherwise have known.

She seemed to fear people, too. If you once became impatient with her she was ever after a trifle cowed in your presence. She could, one felt, read character and detect meanness, dislike, impatience. She could tell at a glance in what mood her master found himself, and by watching her he too could tell in what mood he really was, and feel ashamed when Lassie slunk before him.

Twice a year assorted glands began pumping courage and terrific determination into her veins. This change emboldened her to roam in the countryside for miles around, leaving a broad trail. You should have seen and heard what turned up after that: all the mongrel males that rent the night air with their threats to one another, combined with their declarations of interest in Lassie. She should have been properly ashamed of herself for her poor taste. Walnut Acres became semiannually a seething mass of dogdom, with slinking males slouching into the woods whenever a door opened. The temptation to deplete the canine population was often strong upon us.

We'll never forget Pete. He was one of her early, very small suitors. A physical ailment caused his head to bob up and down perpetually. Pete came from the village across the fields and belonged to the general store there, but daily he made his trek back and forth. He comes to mind in our common family memory as Shaky Pete.

Those were the days in which we fought other dogs' interest in Lassie; we kept her locked in. First the outside boards were pulled from the woodshed—and Lassie had puppies. Next, the males dug under the smokehouse foundation (with Lassie digging her share, too)—and again she had puppies. Then they tore the chicken house door bodily from its hinges—and Lassie had puppies. The last time we chained Lassie in the barn. Our girls stole up gently one evening and opened the door just a bit; they wanted puppies—they got them. Oh Lassie, prodigious puppy producer, farewell.

Up in the woods, under the hemlocks with their lively little cones dangling, where monkey vines twist into swinging, climbing fun for children, there one can find a little flat stone with the inscription OUR CHECKERS, for a little kitty who followed her mother out into the rain one cold evening and fell asleep forever. She was buried tenderly by our little girls; all ribbon-bedecked and paper-wrapped in her little box she was, too. Childhood tears moistened her little grave.

By the side of Checkers now lies Lassie, she too in her box, with house-forced forsythia on top—Lassie, with whom we have lived longer than with any other animal; Lassie, who in her gentle ways showed us more of ourselves than we by ourselves could see. Cheeks were not dry as we tenderly bade her farewell.

Oh, wind from the mountain, as you sweep over the little village and across the fields to the woods, sigh gently through the hemlock boughs over Lassie's head. Speak to her of spring, which already holds a promise before us. Tell her of the sights and sounds and smells that meant so much to her. Thank her for the love and affection that were always hers. And carry her spirit into pleasant fields replete with delectable celestial odors. Peace be with you, Lassie.

DANDELION TIME ❦ MAY 1985

The homely, plebeian, golden dandelion flowers are everywhere just now. We can't seem to win out over them, and so we have joined them. They have earned our respect for their persistence, their cleverness in hugging the ground, their tremendous roots, and their incredible spreading propensity. They surely know their part in the scheme of things. The time for eating their leaves has once more just about gone by. They are tough and overpowering now. But I must tell a little story.

In a long-ago world, a small, impecunious country lad earned a few dollars each spring by seeking out dandelion plants for greens. They were one of the earliest edible growing things, and so after a long, greenless winter, people sought them avidly as a first spring tonic. Whole small potatoes boiled in their jackets, dandelion greens awash with a sweet-and-sour sauce, all topped with sliced, hard-cooked eggs, made a dish fit for a king.

Intense pleasure was afforded by the hunting and eating of these first-ever greens. They took a lot of stooping, but backaches were ignored in the joy of gathering this burgeoning treasure. There it was, everywhere, all one could possibly want.

The lad would hasten to this job immediately after school and a change of clothes. With fingers and toes usually chilled to the bone by the raw spring weather, he would dig up, cut off, and pull apart each plant right in the field. He discarded the brown portions, kept bud and all of the green. By dusk, half-frozen but with cheeks of flame and heart aglow, he would have a twelve-quart bucket full. How pleased his mother would be with the drive that seldom gave up until the job was done. Many bushels would be gathered, cleaned, and sold to the neighbors—a few pennies for a strawberry box full.

Flowers, fragile things at best, speak to us of the true nature of life itself. One rough bulb holds all life within its graceful contours. It is a universe of its own, singlemindedly set to produce beauty, its reason for being. Only sun and rain and soil are needed to draw it forth, to explode the gentle time bomb.

How the dark chambers of the soil must welcome the seeking, sinuous roots, which appear as if by magic. How the root hairs must respond to an overpowering yearning in the direction in which lies life and growth. How delicate must be the searching, the reaching out, the first tentative contact of the roots with the surrounding, supporting life elements.

The soil and its activities, the bulb and its potential, are made for each other. They have their being only to complement each other in the ever-present capacity together to create splendor. Who of us could have begun in our hearts to devise such a driving, made-for-each-other concept and program to keep beauty forever alive?

The flower's color and fragrance are beauties fit for an offering to the deity. No one understands this worth better than the residents of the holy city of Hardwar, in India. Almost half a century has passed since Betty and I first visited Hardwar, whose name means "doorway of God." At this holy city the mighty Ganges River, fresh from the glaciers, bursts forth from the Himalayan foothills onto the plains.

One day we were shepherding a troop of high-school-age people on a tour of the city and its surroundings. Holy men of all degrees of sanctity made Hardwar a center. Bold monkeys lived everywhere, scampered over and among buildings and trees, snatched students' lunches right out of their hands.

Down at the edge of the river were numerous temples, where people came to worship and to bathe. Near the shore one could see hundreds of lovely little floating baskets of flowers beginning to move out into the river proper. These baskets were simple things, each just a bare leaf with sides pinned up by tiny sticks. Each leaf-basket held a bit of fresh water and a cluster of fresh, varicolored flowers. One purchased these at one of the Hindu temples for a few pennies. One then placed them in the rolling water, which carried them out into the main stream, a part of a bobbing, undulating, loosely woven flowery carpet. Each was an offering to the Creator and Preserver of life.

I purchased one, took a quick sniff to smell of the loveliness, then placed it in the water. "Ah, Sahib," said the vendor gently, "you have stolen the perfume. You must not smell it first. It is no longer a fit offering." Disconcerted, yet taught a lifetime lesson, I purchased another offering and left it uninhaled.

A little brass vessel, filled that night with water from that spot in the surging Ganges, still sits on a bookshelf at Walnut Acres, a fitting concomitant for a May reverie.

OF FARM AND FOOD ❦ JULY 1985

Almost forty years ago, when our future was still veiled, a dear friend came from New York City to visit the farm. He felt strongly that the idea of sending foods by mail was weird at best, possibly ill-advised. We had just begun shipping our lovely brown eggs in those early metal shipping containers that were made for parcel-post handling. This notion was almost too much for him. Who would pay all that money for eggs?

Customers would send us their empty cartons whenever they wanted a refill of our really fresh, truly natural eggs. Different-sized cartons held from one to six dozen eggs. The green of the grass pastures added to the yolks a depth of color that is rarely seen today. The flavor was so far better than that of today's battery eggs as to permit of little comparison.

Our hearts bled long ago as they do today at the thought of those dreadfully and pitifully caged battery hens, unnatural from start to finish. They are so rarely seen by consumers, hidden away in long, windowless metal buildings. Cruelly confined, several to each wire cage, scarce able even to turn around in their cramped quarters, their feet having to sprawl out unnaturally over a stiff wire grid through which their droppings fall, they lay their synthetic eggs obediently until they are completely worn out. Enter the soup pots of commerce to deal with the remnants. How soon we forget; out of sight, out of mind.

The tide of egg-shipping cartons ebbed and flowed through the postal system with surprisingly little damage. Each egg was cushioned in its metal shell with pieces of crumpled paper, which normally absorbed the shocks of transit. If fate decreed a rare unhappy journey, we gave credit for each broken egg. Several times monumental messes arrived with a mucilaginous coating spread abroad in the carton, perhaps even uniting the whole into one profound omelet.

Our friend partly accepted the fact, but never, we felt, with full comprehension, that people everywhere were beginning to realize how their fate could be to a large extent in their own hands. They had come to exercise their inborn right to use just plain common sense in deciding how best to take care of their bodies, these temples of God. Gradually they stopped following the herd. No corporation, no high-pressure glossy advertising, no government edict could come between them and their essential being. Their minds traveled to the far edges of the Earth and brought back accounts of living habits among some of the world's healthiest peoples. Here was an area in which they could exercise freedom of intelligent choice, and they did just that. In so doing they helped to change the face of society. They found the truth, and the truth made them free.

The last straw for our long-since departed city friend was our peanut butter. "Who on Earth," he would ask, "will pay to have just plain peanut butter shipped by mail when the market is already so saturated?" Oh, if he were only here today to witness the yearly transmutation of many tons of live peanuts into our simply delicious, fresh, unbeatable butter.

One fine spring day in pretractor years he asked, out of a blue sky, if he could harness and take out to the field one of our teams of horses, all on his own. He had done so elsewhere, he indicated, and was acquainted with all the intricacies. We acquiesced and went to our milking in a neighboring stall. Through the thin dividing wall we could pick up hints of frustration from the stable. Long pauses would follow sessions of apparently fruitless grunting, rustling, and jingling of man, beast, and harness. At times the whole project seemed to be under silent review while breaths were being caught. Finally came the soft, somewhat abashed call for help.

He had not succeeded in fitting the hames, to which the unwieldy harness proper was to be attached, into their proper positions on the collar grooves. Rightly done, this hookup places the burden of pulling the load squarely on the animals' mighty shoulders. One quick glance illumined the inadequacy: the collars were upside down. We all had a great time over that.

Ten or fifteen years later, the large multinational food companies, having first unmercifully ridiculed the truth-seekers' eccentricity, began ever so cautiously to reverse themselves. They came to see that they could not, no matter how valiantly they tried in their advertising, transmute the dross of artificiality and devitalization into the gold of lifegiving completeness demanded by the people, who were just beginning to rejoice in their new-found independence. And so the large companies carefully and gradually changed their approach, testing the waters with each reversal of policy. Unable to hold back the wave of the future, they decided to ride it in. We are reminded of three phases of growth of a new idea. First, people reject it; second, they ridicule it; third, they claim it as their own.

Bit by bit they began to "discover" natural foods. They could make it appear that they themselves were the very first to think of this marvelous new approach to foods. The word *natural* was squeezed and twisted to make it fit new, self-serving definitions. Because we were one of the early and best-known groups in the field at that time, we were approached by several of these large corporations. They wondered if we would be interested in supplying them with natural foods. We did not care either to be compromised or to be swallowed up, and so we did not parley long with them. Nevertheless, good has come out of their approach, and the world is better for their change in direction. Walnut Acres could not feed the whole planet, and half a loaf is better than none. We still have far to go.

LIFE CYCLE OF PEAS 🕷 MAY 1983

Peas come in a flood, breaking through the ground and rising up overnight. They are no sooner out of the soil than the lovely white blossoms appear—and no sooner in blossom than little peas appear, and then harvest time is here. Those sixty-five days from planting to picking go so fast.

Peas are our earliest vegetable crop. Sometimes in April, when the ground is being prepared, it is bitterly cold. We've seen times when we've had to wear layer upon layer of everything we could put on just to keep from freezing as we pranced our iron steed across the fields by day and night. In April one must spend in the field every minute that the ground is dry enough to "work," lest another showerfall make one do it all over again. The weather squeezes us in spring.

Let the ground temperature four inches below the surface be 45 degrees Fahrenheit and you'll have the warmth that peas love most. They grow like mad, their tendrils grasping each other in loving embrace until the whole field is one continuous mat of intertwined pea vines.

A breeze blowing across the pea field sets up a gentle motion. The lovely clusters of filling pods, in rows seven inches apart and suspended a foot or so above the ground, sway softly under their shady green canopy. As they fill out, the springlike tendrils are stretched, gravity drawing the heavy pods lower and lower. Then one day in June, along comes the cutting knife. A six-foot swath is cut and rolled into a windrow. A green-crop harvester picks up this row and lifts peas, pods, vines, and all into a truck. From here they are placed in a gigantic, clumsy, ancient peaviner, which devours stalk and all, aborts the peas from the pods, and deposits them in containers to be taken to the cannery or freezery.

The pealess haulm is spewed forth at the other end of the pea-monster, to be placed in huge piles, that part of its work done. There, through Nature's wondrous alchemy, the plant loses its identity, its life as a peavine, only to become a butterfly of hope as it is transformed into rich humus to feed the soil, another link in an endless life-and-death struggle.

NOTHING EVER CHANGES ❦ APRIL 1976

On May 29, 1453, the thousand-year walls of the Jewel of the East, Constantinople, were breached by the Turks. The city fell in ruins. David Dereksen, in *The Crescent and the Cross,* reports of that time, "Constantinople is noted for its fireflies. That night [after the fall] they flitted and wavered as usual, through the abandoned gardens and open spaces of the city, around flowering shrubbery gone wild—for it was late spring—twinkling in and out of sight, no higher than your waist. *For nothing ever changes. It is we who go.*"

Last year's mallard couple, one with a powder puff tilted jauntily on his head, raised a splendid brood, some puff-headed, some not. They all lived together happily during summer, fall, and winter, in joyous cheeping and quacking communion. Now suddenly we notice that the old couple has gone off on its own again, having sought out a hidden nesting place by the pond. One small duck egg, harbinger of more and larger to come, has already appeared. Nothing ever changes.

NIGHTTIME IN THE BARN ❦ MARCH 1982

What memories our old frame barn could bring forth, were it a talking creature. This mammoth structure, with its hand-hewn, oak-pinned beams, its spacious hay mows, its cool bank-backed lower level, could tell many an engrossing story. Of lambs, for example, seeing the first dark of night, long before dawn in the bitter-chill frostiness of early February.

To aid in the birth of a lamb, one hunkers down in the sheep-pen straw, surrounded by warm, woolly, lanolin-scented closeness, awaiting the final push that transfers the lamb from amniotic warmth to the frigid strangeness of new life. Sometimes, instead of appearing normally, nose first, the lamb's head is turned back. The whole birth procedure must then be delayed until the little fellow can be pushed back, to be straightened out by the shepherd's hand in the birth canal inside its laboring mother. Without this help, both could be lost.

While waiting for distress signals, head nodding in the dim, mysteriously flickering light of the kerosene lantern, one hears a concert of night sounds from the light-untouched recesses of the stables and mangers. Horses stamp on the hard clay floor as they tug voraciously at mouthfuls of the hay stuffed into their overhead racks. They munch mightily on the dry timothy stalks, reducing them in short order to mush fit for digestion. They huff, puff, and snort, making sure their presence is not forgotten. An occasional whinny speaks of their detecting the shepherd's presence.

Cattle low (speak lowly) to their infant offspring (hummies), who in turn hum sleepily. Sucking, lip-smacking noises tell of nourishment being taken at all hours of the night. How human-sounding it all is. Cud chewing by the ruminants goes on all about one, as the food swallowed earlier in haste is brought up from the first stomach for a more leisurely, thorough treatment. When the food is finally prepared for its journey to the second stomach and beyond, the animal's head goes up. In the dimness one sees the outline of the finished lump descending. Soon thereafter another cud is cast up, to be chewed in its turn.

A sleeping cat hunches nearby, hoping subconsciously, even at this unusual hour, for a squirt or two of eye-and-nose-filling milk direct from the source. Nothing could be fresher. It must be worth almost strangling, because that powerful stream covers the entire face. With practice, and with the milker cooperating, it's not hard for the cat to become adept at opening its mouth really wide, and with frantic gulping to imbibe a high percentage of that sweet, warm stuff of life. But all that force really does tickle the cat's throat, and much sneezing follows every lunch-squirt. (It takes so long to lap milk from a saucer.)

The old farm dog snores away in the sheepfold corner, his paws curling and uncurling in response to events in some other world. Now and then a particularly hard leg jerk awakens him just enough to register that all is as before. A tiny scurrying in the feed trough betrays the mouse searching for missed grains. A sleepy chicken, roosting among the cobwebbed logs overhead, bewails its lantern-inspired disturbance with a couple of miniature squawks, then slips its head back under its warm chicken-down wing covers.

At last the vigil draws to a close. After numerous tries and much straining, a shiny, wet black nose appears. Then eyes, forehead, and ears. Sometimes a ewe will walk around the fold with a lamb's head exposed sternwise. The little fellow will be viewing the transient scene with interest, resting literally in two worlds. Then finally, sometimes assisted by the shepherd, the whole compressed body appears with a slippery plop, and a new, tiny, plaintive voice is heard in the universe. The lamb is unhooked from its mother, patiently suffers washing and drying and motherly bleating and nuzzling, and wonders where on Earth it is. After a few minutes it tries to rise. What wobbliness! With many a haunch-sitting and forward-pitching, it finally gets its nose near the mother's woolly flank. What a job to find a small teat in all that woolly forest. The lamb begins to nose and suck here and there. Sometimes the shepherd must let the lamb suck his finger, then try to slip the engorged teat into the lamb's mouth when it has gotten the idea. Some trying, struggling moments are accompanied by thrashing about and muttering. But finally peace and abundance reign.

Sometimes the mother won't let the lamb suck, and so it becomes a bottle-fed house lamb, bouncing stiff-leggedly between barn and house, accompanied by joyous little anticipatory bleats. The children love welcoming this visitor. Often twins are born, and once in a while triplets.

The shepherd-farmer, drawing and holding together this incredible barn community, knows each beast by heart. The sounds the animal makes, its idiosyncrasies, its way of thinking are all part of the farmer's understanding. He anticipates its every move and mood, appreciates its individuality. He becomes a part of his barn family, feels himself just one member of the mutual support system that ties all into a glorious whole. Truly, this is a complete world of individuals, beautiful in its understanding, accepting in its appreciation one of another, each doing its own thing successfully, untroubled and uncomplaining. This relationship asks for little and gives much. Where has all this beauty gone, in our hunger for production, speed, efficiency?

SPRINGTIME IS FOR LAMBS ❦ APRIL 1958

How I wish you could see our lambs. Few animals are cuter. We realize it now afresh. We so easily forget during the rest of the year, for lambs are so soon sheep, obstinate and offish. But now, bless their little hearts, they are still loving and trusting and cuddly, and bouncy with the sheer joy of being alive. We have never found anyone able to resist them.

Three ewes have reproduced thus far, giving us five lambs. (A neighboring farmer's ewe gave quadruplets, all living.) There are male twins, female twins, and one husky boy lamb. What a time we have been having, too. We've had to begin two of them on a bottle and nipple, for ewes sometimes refuse to own and nurse lambs. One of these two will have to be raised on the bottle entirely. She has become, willy-nilly, the household pet.

When evenings and nights are cold, the kitchen and a corner of the cellar are hers. She is not housebroken. From the time she is brought to the house she is the center of things. She bleats plaintively, butting everything within reach, striving to extract milk from it, until her formula (cow's milk with a little honey) is ready. Then her little body becomes a wriggling, squirming mass of lambskin, held up by prancing sailor-pants legs. Every four hours, eight ounces of milk are thus transferred into a most appreciative lamb stomach. When later her sides are again undistended, she is ready anew.

Between meals she entertains. She has usurped one of the hot-air registers, much to the disgust of the hound, Wendy, who reckons that it is her own. The other dog—little ragamuffin Tinker—loathes her with his whole being. With ill-concealed jealousy he lunges at her, nipping her nose if he thinks himself unobserved by human beings. Whenever we show her any affection, from his retreat under the table he rolls his eyes in torment. She remains blissfully unaware of the tensions her presence sets up, even daring to sniff at his feed dish—which is, of course, the last straw. The ensuing scene is not calculated, along with all the other goings-on, to keep the kitchen a model of orderliness.

When placed in the sheepfold with the other lambs and sheep, this lamb is not at first sure: Is she sheep or human? Confusion is in every feature as she stands between us and her family. Then suddenly, with a stiff-legged hop, a rascally shake of the head, she joins the sport of the lamb troupe as the others race through feed troughs, climb upon their mother's backs, and butt heads with fierce glowerings.

And still she feels the pull of a voice beyond herself and her peers. Her deprivation has unwittingly opened for her a door into a wider world. For at my call, like iron drawn by a magnet, she separates her little body from the common woolly mass, crossing with unutterably lonely, trembling steps through a no-man's-land, against the unbelieving stares of her kind, until she is safe in my arms. I like to think this is a matter of more than just milk. And her helplessness, her response to my call, fills a need in me as well.

For each of us—for every living thing—each spring may be the last spring in which to try our powers and our faith. Perhaps many such seasons remain, but there will always be one fewer. For everything but life itself, that is. And because life does go on unceasingly, and because we grow most when we contribute most to that continuation, and because for us springtimes grow ever fewer, we poor-rich thinking creatures must never fail to learn from those life forms that are moved only by instinct or less. Surely our faith in spring should exceed theirs.

AMISH FARM SALE ❦ APRIL 1978

Last Saturday I attended the first farm sale of the year, to purchase an old farm wagon for hauling in our tomatoes this fall. An Amish family was selling out, preparing to move to Paraguay, where they will have room to breathe, where life is simpler, and where one may still pioneer. The snow was melting from the hills above the house and barn, and every step was a battle with inches of squishy, slippery brown glue.

Theirs was a hill farm, with view dropping off into rolling, snow-covered dunes to east, south, and west, an eye-filling bit of paradise. One could not see another dwelling anywhere. Though fifteen miles from the nearest town, the family could feel that, in time, insidious sophistication might destroy its way of life. One wonders what will happen to such spirits when the last square inch is conquered and there is simply nowhere else to go.

One sloping field a thousand feet from the barn had been turned into a parking lot for horses hitched to their black Amish buggies. There they stood, the beasts shifting from one foot to another, hour after hour, waiting quietly through the whole day for their masters' return. Half a hundred of these buggies stood in stark outline on the eastern horizon. Standing on the barn floor behind the auctioneer, one could see these buggies framing the closer-in view of the crowd.

What a bazaar! Amish and Mennonite families were predominant, different groups or orders in garbs of varying colors and styles. Broad black hats and dark-colored coats and trousers for the bearded men were invariable. The women were bonnetted variously, and fully covered in black or purple or dark blue dresses of one sort or another—all homemade, of course. Small children kept quietly to the vicinity of mother and siblings. One heard no tantrums, no quarrels, no self-assertions. When the mother spoke softly in her German patois, the children moved uncomplainingly to carry out her wishes. She was the Mutter; she knew best. They were just the chicks. Their behavior seemed the most natural thing in the world, and somehow so right. Some of the little ones stood all day, quietly, without question or discontent. That was what one did, that's all. Mama and Papa are all; they know everything.

The parents made a study in contentment and goodness and patience and serenity. Faces of incredible beauty broke readily into sober smiles. Quick-witted repartee showed quiet joy in every little happening, taking in the whole event and responding gently and with wisdom.

The plain, bare house was a nest of mothers and children, some eyeing the "English" people (as they call us non-Amish folks) in curiosity. It was a day of no work, hence quite a picnic. One four-year-old boy, sprawled fast asleep on the couch in his big black hat and coat, would have torn at the heart of any grandparent. Our various worlds were one for a time, and we all grew bigger and better because of it.

At the sale were hundreds of farmers in work clothes, a sea of farm faces, "English" and Amish and Mennonite, keeping up with an auctioneer who could see everywhere at once. These farmers are the people whose job it is to keep life in the bodies of their fellows. One could sense deep-rooted uneasiness over their own situation. They can no longer buy the new farm machinery—who has thirty thousand dollars for a new tractor or combine? What can they do as their machines wear out? Where is the end of it all, as they work harder and harder for less and less? We were told that six hundred farms within a hundred-mile radius had reached the point of no return and could go under at any time. The trend is both tragic and dangerous. We farmers all do the same thing: we put patch upon patch. Here at Walnut Acres we have eight tractors, most of them between twenty and thirty years old. No one ever buys anything new nowadays. Life is a constant, daily going backward. One wonders what the outcome will be.

The day ended in a chat with a dear Amish friend who has six or eight of the most beautiful children you ever saw anywhere, all delivered by him and his wife in concert. It has been a deep joy to work for twenty years or so with this bearded, black-hatted philosopher and to have him raise a few things for us from time to time. I learned that now he too is leaving. It's getting too crowded; land is too scarce and expensive. A hundred-acre farm, bare, now sells for up to $150,000. Ten families in his church have purchased 800 acres in the lower-priced, uncrowded Kentucky hills to start all over again. There they will build their houses and their barns. It will be their third move since we have known them. We shall be far poorer for their going. The spirit that made our country great: may it never be lost under our wealth and security.

In leaving, as I drove down the hill into the more humdrum day,

I passed a shiny, black open buggy in which blossomed two red-cheeked late-teen-age girls. They were just then sharing something that lit up their eyes and faces into seraphic beauty. Many thoughts accompanied me the rest of the way home.

DREAMING ABOUT TREES ❦ APRIL 1959

We planted about fifteen hundred small trees again this year, some evergreens, some deciduous. It's such fun to dream about trees. Fifty or a hundred years from now, will anyone know or care by whom they were planted? Someone, sometime, is sure to notice the straight rows and wonder about them momentarily. Even if someone doesn't, the planting is its own reward. I hope young lovers run laughing among them sometimes, or swing along hand in hand, faces lifted and shining, dreaming glorious dreams. One of the greatest things we can do is to provide conditions in which future generations can do their dreaming. Perhaps we will even learn some day that life lies more in the dream than in the reality.

RALPH FALLS IN LOVE ❦ APRIL 1966

As we and spring prepare to enter into one another, I must tell you a tale of romance that has developed over the winter; I cannot keep it to myself.

Perhaps twelve years ago, a tiny, homeless, ragged ball of hair was found wandering, unimpressed, in front of the Indian Embassy in that holy of holies, Washington, D.C. Out of pity, the Animal Rescue League impounded him and kept him in its dogormitory. How crushing to a little blithe-spirited wanderer to be caged, even in ideal surroundings. What has fate, really, in store for one such unhappy creature?

But to the rescue! A noble-spirited gentleman, ever about doing good, sensitive to the needs of three growing children on a farm in the upper reaches of Appalachia, spotted this small scamp. Moreover, utterly taken, he saw to the details of expounding, naming, and importing him into Walnut Acres. The suggested name, Guggenheim, was rather discourteously scrapped for Tinker Bell. Those were Peter Pan days, you see. And then further, of course, the name went to just plain Tinker.

Life has been more than good to Tinker. Growing old along with him, the best yet to be, is the elderly spayed black hound, Wendy. What a pair! No rabbit is safe from their fierce barkings. No woodchuck may stray unwarily more than a mile from the safety of his burrow. No thief dare risk crashing down unfamiliar steps to land at their noses. A house-centered circle, a half-mile in radius, replete with brooks and downs and sun-spotted glades, is theirs to command. And guess who snores closest to the fireplace when the chill winds of winter sweep the wild ones into their dens?

Oh, we have seen occasional sneakings off, under strange urgings, for Tinker to father most peculiar-looking, later completely spurned pups. And it is very hard indeed for a four-legged puff of skin-wrapped life to remember that manure piles are *not* for rolling in, and that weed-seedpods just love to cling to long, furry hair. But with all its ups and downs, life has been good for a friendly, inquisitive little fellow.

Well, up until last Easter, that is. Then Ralph the duckling joined the family. Oh, it was all harmless enough. A one-day-old ball of yellow duck down, demanding, untrained, but sweet, is nothing to get excited about. Until suddenly it is ten or eleven months old, is full grown, and has stolen an unjust share of human attention and affection, of which one can find only so much. Even that can be borne if necessary. But not this latest.

There's just no other way of saying it: Ralph the duck has fallen in love with Tinker the dog. Let Tinker show himself on the bank and Ralph comes dashing out of the most tempting pool. Where Tinker lies in the sun, there sits Ralph close by his side. Let Tinker stand up, thinking to go elsewhere, and Ralph stands up also, thinking to go along. It appears a Platonic enough relationship. Only once in a while will Ralph nuzzle Tinker with his yellow beak. Generally he seems content just to be in Tinker's dear presence, beak to nose, so to speak, staring soulfully into Tinker's violet orbs.

Tinker is not only uninterested, he is positively revolted. Disgust is obvious on his face. Sometimes his feet seem shod with lead, for when he moves, Ralph moves also, and he is in a state of perpetual checkmate. Tinker has adopted all sorts of subterfuges to keep out of Ralph's squint-eyed sight. But love sees around corners, reads all signs as favorable, feeds on contempt and disdain. Tinker does not always take it sitting down. His lip curls, he emits blood-curdling yips, and we look outside to see Ralph's long white neck in Tinker's mouth, feathers flying, Ralph a-flapping. But when things have cooled down a bit, there stands Ralph, right by Tinker's side again, more certain than ever of a place in his affections.

We haven't yet consulted a psychologist about this affair. But should Tinker one heavenly spring morning be missing, we're going to head first thing for the Indian Embassy. After all, one can take just so much.

Summer

Nothing is busier than a bee, or more tenacious than a daisy, or more inspiring to grow and harvest than a tomato. These summer pleasures embody the mystery of life —something one may touch even more closely if one tries dowsing.

MOVING DAY FOR THE BEES ❧ JUNE 1949

Bees, so common, are so highly organized, so complex. Let a handful of these animated, diminutive airplanes, engines droning, show more than passing interest in even the strongest man, and he will probably bow out as gracefully as possible, remembering something he has to do elsewhere. They don't really face much opposition, except maybe from Winnie the Pooh. Someone who wanted us to consider selling his honey told of the trouble he had with bears coming out of the nearby woods and tearing his apiary to pieces, apparently oblivious to the disapproval showered upon them by the defenders' kamikaze attacks. How many bees must expire in this slaughter of the innocents, as they leave their stingers in the bears' less shaggy parts, only to have them at the same time torn from their very own abdomens as they fly away, eviscerated, to die.

When bees are happy they want nothing more than to make babies and honey. The queen must be the happiest. Bred for a lifetime in her nuptial flight, she can lay an unbelievable number of fertile eggs. She pastes each one fast to the bottom of its comb-cell at precisely the same angle as the next. They look like marching soldiers in serried ranks. Then the infertile female workers take over. They see to it that these eggs turn into more infertile female bees. After a spell, out of each cell crawls a fat, buzzing baby bee. How engaging they are as they practice flying, at first sticking close to the hive like a tetherless astronaut to his space home.

Let it be too hot, too crowded, too unexciting, and the bees will start making queen cells, like peanuts stuck to the sides and bottoms of the honeycomb frames. The queen blithely lays an egg in every

cell that comes along, even if it means she is creating a deadly
enemy queen. (Hers but to do or die.) Well, soon too may queens
are around; something must give. And so the hive splits up. A mass
of bees decides to leave when their favorite queen wants to go.
What a dither. You'll see and hear them like a living cyclone,
whooshing across the fields to a likely spot picked out earlier by
scouts. Sometimes they go far, sometimes near.

How very busy the bees are just now, in spring. They are driven
creatures. Away back in the early years we had a number of bee-
hives. We learned much about them by loving them as much as one
can love sensitive, defensive, prickly objects. In those days we were
still learning by the books. Any new, unaccustomed thing that happened was cause for us to rush helter-skelter for the appropriate volume to find the answers.

One warm day in the springtime of both life and that year, the
telltale signs appeared. A great busyness, a dashing hither and yon
of thousands of bees near one of the hives told the story: Madam
Queen was about to leave the hive, having had all she could take of
her old, cluttered house. All she wanted was a new place. She had
to leave even if she didn't know just where she would be living
next. And soon out she went, as we watched, utterly fascinated. As
her tremendous body, appropriately dignified, left the hive and became airborne, she left a trail like a giant vacuum, sucking thousands of delirious worker bees from hidden reaches in the hives.
That amorphous mass of darting, dashing, dauntless creatures
moved ponderously through the air. Each bee seemed to want to be
as close as possible to its queen. Streamers of latecomers trailed behind the main cluster.

Instead of flying far, this time they flew high. Almost straight at
us, in fact, went the queen, with her retinue clawing each other
madly all around her, settling droopily on a black-walnut limb thirty
feet above the ground. Out came the beekeeping book. Out came a
basket in all readiness with a long rope tied to the handle. Out came
a ladder. Up the tree I scrambled while my helper read aloud the
step-by-step procedure.

Precariously, I reached the heaving, ecstatic swarm. Holding the basket under it, I gave the limb a mighty thwack. Down fell the queen and her disciples into the basket, with hundreds of unhinged individuals zipping bewildered around me. Down went the basket on the rope. Down went the beekeeper, to dump the container in front of a fresh hivebody containing a frame or so of eggs and brood. Suddenly very docile, into the hive walked the queen, followed by her uncounted subjects, and life began afresh. It had worked: we laughed rejoicing as we realized that we had done it. We had coped—that is, with the help of the book, the basket, and the bees.

THE OLD SPRING HOUSE ❧ MAY 1983

In the early years on this farm the oldtimers needed a spring house to cool the milk, and to serve as a refrigerator in general. The iceman probably did not get out to the farms very often, and indeed with 40-to-55-degree spring water, one could manage very well without him. The farmer had to have water deep enough to reach to the neck of his milk cans, each holding ten gallons. If he had a mess of cows, he needed lots of spring-house space.

Our old spring house is just by the side of the hundred-year-old farm lane, which leads from the entrance road to house and barn and other buildings. The burbling stream runs directly by the house, between the narrow lane and the fields. The never-failing spring, which supplies the house with some of the world's best water, lies on the field side, just beyond the stream. Its own little house is built over it, to restrain small beasties from too much sharing with us persnickety human beings. In return, the wildlife may have the whole of the overflow that runs from the spring house to the stream, you see. The stream ultimately gurgles its way into Penns Creek, perhaps three hundred yards below the house as the stream meanders.

For the dairy farmer, the spring house was an absolute necessity; cows' milk had to be kept cold. In those days the unsophisticated cows were all milked by hand, of course, and what a slave the farmer was to those animals. We think of them as his cows, when in actuality he was their man. He was at their every moo and bellow, at all hours of day and night.

In the evening the cows were milked at a set time, regular as clockwork. The milk was poured into the milk can from the milk pail. On occasion, the cow, in a restive, impulsive moment, would set her foot down precisely in the center of the full bucket. One hesitates to remonstrate too vociferously with a nervous cow while sitting cheek to flank, a hair's breadth from a powerful kicking machine. Would that the udder were elsewhere. The best one could do was to push with might and main into her flank section with one's head, making it almost impossible for her to spring into action. Of course, she could simply move away and *then* kick. The proceeding at best left one's coiffure unbelievably fragrant.

Often a water-filled trough in the barn held the full cans until milking was done. Then they would be loaded on a cart or wagon and taken to the spring house. Ours was a hundred yards from the barn, down behind the house, built over another spring, separate from the one feeding the house.

Here the milk cooled overnight. But the farmer had to be up in the wee hours, both to milk the cows again and to cool the milk before the milk collector came in his truck; the collector would complain if the milk was not really cool. The milk receiving station would not accept warm milk because of the danger of bacterial growth. On each trip the collector had to visit a set number of farms, hoist the cans full of milk, take them to the milk plant, and bring back washed, empty cans. Everyone hustled to meet a set schedule. The job built men. We recall one short-statured fellow who had the knack of throwing those hundred-pound cans of milk high onto his truck, day after heaving day, and making it look easy.

Among farmers and those who worked with them a camaraderie helped make it all seem worthwhile and aided in spreading community news from farm to farm, brightening the day. Farmers did what they had to. If you had twenty-five cows, you got up earlier than the person with twenty cows. Muscles of hand and forearm grew hard as iron from emptying a hundred teats twice a day. Now, with milking machines, the job is easier. Today dairy farmers often milk thrice a day, at eight-hour intervals. They get more milk in that way. No wonder there are so many empty dairy farms in New York state and elsewhere, the barns sliding back into the earth from which they sprang. Well, that old log spring house at Walnut Acres in which the milk cans were cooled has long since been elevated to higher things. It is now a hideaway where one can be out of the world's reach. It was carried to its new location behind the woods. Here are no springs, no gurgling streams, but weasels and mice and squirrels occasionally find it an agreeable place to investigate, although they find no food there. Hornets and wasps, bumblebees and honeybees love to bore into its log shell. As I sit there on warm days to meditate and to write, I become a part of a buzzing universe with countless insect wings beating the air into marvelous hums and drones. In the wintertime it is semi-retired from its many lives. What stories it could tell of the early days when it was the dwelling place for a pioneering family.

LESSON OF THE DAISY 🦋 JULY 1983

Daisies in unwanted areas may be weeds. They will grow in the most unlikely places—along railroad tracks, on steep banks, in wet fields. Yet how thoughtful of Nature to endow us with such sturdy, persistent, indomitable examples of beauty created from what one finds at hand.

Have you ever tried to pull or dig up a fair-sized daisy clump, to plant in the back yard at home? One's respect and insight grow with the digging. This wondrous plant makes the very most of circumstances that may seem poor to us. Those copious roots spread in all directions in an incredibly thick, interwoven mat. As if by determined seeking they have found life.

Years and years ago, a University of Colorado professor named Glenn Wakeham performed an experiment with common string beans. He grew some in deep, rich, commercially treated, fertile soil in a lovely flat, tended valley. The yield of beans was as heavy as one would expect; the crop was good to behold. On the scraggly hillside above the valley he planted the same type of beans in the normally untended, rough, stony earth, in a plot we might call daisy soil. Here the plants were not as lush or as promising as the others. The yield was not as large, although the beans themselves were likewise good to see. But upon comparing the nutritional analyses of the two crops of beans, to his surprise he found that the beans from the "poor" hillside considerably outdid those from the "good" flat ground in overall quality. They proved to be higher in minerals and other qualities that count than their more sophisticated brethren in the valley.

Whether or not this contradiction is universally true, we cannot say. We do know that on our stony hill farms, where we have used good organic methods for many years, the crops do far better than we expected them to. We had to pull out many trees that had taken over fields used earlier for crops. By farming on the contour in narrow strips, alternating sod crops with plowed and planted field crops, we have been able to heal the gullies worn during centuries of subjection to the old method: horse plowing up and down the hills. By putting back far more than we take out, in animal and green manures, in chopped and plowed grasses and legumes and heavy sods, we have fed earth's stomach, the topsoil, a completely natural diet to its own liking. We've fed the soil and let it feed the plants. And we've seen yields increase apace.

Leave soils undisturbed for several years and some daisies may come back. I am not equipped for Professor Wakeham's complicated tests, but I am content with the belief that, by using our completely natural methods and no synthetic plant foods or poisons, we are bringing out the best of the possibilities in difficult conditions, to the ultimate benefit of every living creature connected with the land.

Perhaps in the poorest places, conditions, and persons reside basic possibilities, awaiting the proper care and treatment to bring out their best, just as the white and gold of the stiff-stemmed daisies reflect their great power to find hope and comfort in possibilities that lie all about them, in circumstances that, in our short-sightedness, we think under par.

Whenever daisy time comes around once more, this grand concept sweeps over me. The perfection of the soft petals growing from the golden center, the grace and beauty of their form and attachment, the joy of life they exhibit, all betray the sweet secret. Life and power and beauty are where you find them. Nothing is too much work in letting one's roots seek out one's life sustenance. Nil desperandum: never despair. Take what you have and build beauty from it. It is possible; it does work. In times when the heart is slow, one must remember the daisies of the field.

We taught for a time in India. Years ago Mohandas K. Gandhi, at his own school there, where I lived for a short time, said, "You westerners come to India with your schools, to educate our children. You give them electric lights, running water, tables, chairs, cutlery, beds, desks, and so on. When they leave your schools, they no longer fit into village life and must go to the cities, where they may or may not survive. In our school, we give them no more than they have at home in their poor villages. But we teach them what they can do with what they have there. This is real education."

We have a color photo of Malina, our precious flaxen-haired granddaughter, almost lost in a sea of sprightly, burgeoning daisies. How we hope that she will learn to take a leaf or two from their book. Let her be content with what she has. Let her roots go deep into life's soil, drawing riches from even thin, stony pathways. Let her learn to create beauty out of whatever substance her roots can reach. Let Earth's children, early and universally, learn the lesson of the simple, unsophisticated daisy.

REMEMBERING 🌿 JULY 1982

Many have written me about their early days on a farm. My own stories seem to awaken old memories in people, who tell me all about their golden past on the soil.

Everyone know that it wasn't all roses. There hard work was endless, with everything needing to be done all at once all the time in spring and summer. We had no television, much less time for imbibing its culture. Life was full of real situations that Hollywood attempts at portraying country life make as unnatural as today's artificial foods. One lived out the joys, the triumphs, the difficulties at first hand, and had neither time nor stomach to sit passively inert, absorbing contrived inanities as if they were the real thing.

Who can ever forget when hay was taken in loose, in massive loads? Remember how you had to build up the enormous wagonful in such a way as to tie it all in, so that no section could slide off, threatening the security of the whole load? Remember how terribly hot it could be, when you were on the wagon, in front of the loader, when a thunderstorm threatened? You just had to get the hay in before the rain came. Mounds of sweet, slippery, tangled fodder kept pouring from the top of the loader, just about overwhelming you as you raced wildly to spread it evenly to all parts of a wagon that grew longer and wider as the hours wore on. How those marvelous steady, enormous Belgian horses entered into the spirit of the endeavor, sensing the need for haste.

Once, on our farm, after just getting in the last load at the last moment, men and beasts stood on the barn floor, watching the wind-whipping thunderous clouds swirling eerily about. Someone, brow still aglisten, was heard to shout, "Send 'er down, dear Lord!" Whereupon, in the same instant, with a loud clap and a tremendous gushing fall of water, he was accommodated. The cheers from those assembled almost outdid the cataract's roar. Nothing is so refreshing at a time like that as the washed summer air. And it didn't take long to learn where the old tin roof needed to be nailed down again.

How fast the summer days passed for the children between chores. Remember trying to find where that old white hen had hidden her nest? Or where the bantam family, never penned up, always pert, cocky, and self-important, spent the night? Did you every try to catch a guinea fowl during the day? Could you locate tiny guinea chicks when the mother was leading them foolishly through tall, wet grass early of a chill morning? They would be cheeping all about your feet, but you could never spot the critters.

And how delicious to climb up to the very top of the hand-hewn, hand-pegged beams, to jump gloriously down through the scented air, to land ten or even fifteen feet below in real, natural, resilient, scented softness. How it warmed the heart-cockles of the older children to be rewarded by admiring murmurs and glances from their less daring younger friends and siblings. As for the showing off when the city cousins, prudently discreet, came for a visit, that was absolutely shameful.

Does anyone recall that old poem by James Whitcomb Riley, the one that began,

Blessings on thee, little man,
Barefoot boy with cheek of tan,
With thy turned up pantaloons,
And thy merry, whistled tunes—
Outward sunshine, inward joy,
Blessings on thee, barefoot boy.

How wealthy we were in those days, when we shared with the Amish the richness of not wanting much.

Farm children miss much that their city cousins may experience. Only time will tell if all they gain from the soil and the seasons is in reality a part of the heritage of the race that cannot, may not be safely despised. We hear the old folks remembering, the young wanting to learn. From the soil we came, by it are we sustained, to it we return. The soil enters unto us at so many points that we may fancy our very blood is liquid soil. One wonders if a mystical homing bond connects us and earth, controlling more of our thoughts, our emotions, our longings than we are able to—or care to—recognize. At least the physical part of us is but walking, talking earth. We would be wise to treat our bodies as wondrous expressions of the same life.

Do you remember the myth about Hercules in his struggle with Antaeus, a sturdy opponent? Someone told Hercules that Antaeus could be subdued only by lifting both feet from the earth, that Antaeus would become powerless as soon as he lost touch. Well, Hercules did, and Antaeus did, and maybe the Greeks knew something that must ever be learned afresh. Parents must teach children respect for and love of earth, for in its strength earth abides. It persists, endures, is steadfast. Living at one with it, we are enabled to mount up on wings, as eagles.

SUSIE'S LIFE ❦ JULY 1986

For his twelfth birthday one of our four grandsons, Nate, had his dearest wish fulfilled. From a nearby small kennel the family brought home the cutest little flop-eared beagle ever created. Oh, they have Meg, the home-loving, smiling collie who greets all and sundry politely. But she isn't especially cute these days. She could be a grandmother with a number of "greats" preceding that title. She is quite settled and sedate. Long gone are the days of puppyhood.

Little squirming Susie just wriggled her winsome ways into the hearts of everyone, including cats, Peter Rabbit, Meg, Queenie the pony, and of course children and adults. This proceeding improved noticeably once she had learned to bathe outside of the house.

For Susie no shoes, boots, mittens, bathing suits, towels, or socks were sacrosanct. She simply wanted to chew them all, carry them around, and in general worry them, all with great gusto.

Outside, however, Susie was more hesitant. Let someone walk over to Susie's home, and out she would dash from her box, but tentatively. It was almost as if she were tethered by the elastic bands of acquaintance and memory to the back porch and her cushion. She would wander just so far from home, then turn her little flexible form ever so sweetly to stare back through the days of earlier puppyhood to something she knew and understood well, her infant security blanket. Soon, her basic baby fears overcome, she would be all over one's feet, climbing up one's legs. Suddenly she would find herself swooped from the earth and cuddled against one's bosom, an alternately snuggling and wiggling sausage of incredible cuteness and affection.

She soon awoke Meg's maternal instincts, brought back some of her youthful playfulness. By the house, the huge, furry collie and the shiny, smooth white-black-brown little pup would roll on the ground together, snarling fearsomely. They had the very time of their lives. Heavenly were the surrounding fields in summer and fall. How Meg and Susie loved to go roaming over the hundred and more acres of which they and their home were the center.

At first Susie trod hesitantly beyond her narrow world, looking back often over familiar terrain before tempting fate with wider forays. She would stay pretty close to Meg, who moved slowly and carefully through the fields immediately adjacent to her home. Meg was not a roamer, which is more than can generally be said about beagles.

Later, as friendly local dogs passed by on their visitations, sniffing their tortuous paths nose to ground through the hinterland of Susie's experience, she would tag along on ever longer treks. It is amazing to notice how very soon a new dog on the block belongs. With no fussing, no fighting, Susie was very soon a lifelong member of the club. Once or twice she got lost when the older dogs moved too far too fast. Then it took a lot of puppyish yipping and crying to get the personal attention needed to rescue her. Thus did her horizons expand. Thus too did her dreams grow sweeter as distance lent enchantment.

The old barn was a place of special interest to Susie. Her cold little nose was drawn as by a magnet to every delicious odor it could detect. The scents that interest little hound dogs must have been sniffed right up to her brain in all their overlapping glory. They must have told her at one time or another of rats and mice, of squirrels and chipmunks, of cats and other dogs, of horses and steers. How far back do scents go for dogs?

Perhaps she detected the passing through the barn of more exotic smells: those of a possum or a coon, a woodchuck or a muskrat, a porcupine or a skunk. Who knows what nocturnal critters explored the open barn and laid down their tell-tale trails for later pup sniffing? Barn snuffling was surely near the top of Susie's list of preferred occupations. Things catchable were normally intercepted by the cats, and so all Susie had were the delectable odors that wove themselves somehow into her paw-jerking dreams. One wonders if at such times her feet got tangled in the covers as she attempted to escape a dire fate or to chase a desperate prey always just out of reach. Susie had a way with cats. Though not overwhelmed with enjoyment, the large, sophisticated male cats, with surprising generosity, suffered her to maul them about a bit. They seemed almost to like her, would rub against her at times. The four kittens also, when Peaches their mother was not about, would act playful and enter into games with Susie. As for Peaches herself, she left no one in doubt as to how she felt about that enemy cavorting with her babies. Susie, after receiving a number of smart swats on her tender nose from Peaches's open claws, learned her lesson. She always found a place to be where Peaches was not when they were both in the same vicinity. Kittens or no kittens, Peaches would deliberately seek out Susie's nose just to keep in good training.

And then came Peter. From somewhere appeared this large, young, whitish, domesticated rabbit. The boys fixed up a cage for him, but it grew so unbearably hot in summer that Peter nearly expired, and so they decided to let him run loose. Soon Susie the rabbit hound and Peter the bunny had struck up a playful friendship. They would chase each other, jump over each other, nuzzle each other. Neighborhood dogs of all stripes, in their wanderings through the farm, seemed to respect Peter as one of the family. To this day they have not yielded to temptation, if such it be.

One day Susie was seen astride Peter, who crouched on the ground beneath his playmate. They seemed frozen for a moment in eternity. Who knows what strange confused feelings reached out of ages past at that moment to shake their trembling frames? Were they being told something they did not want to hear? Were their inherited nerve patterns beginning to border on the adult fear-and-conquest syndrome into which they were born? How long could the hare and the pup lie down together, fulfilling Isaiah's vision about the end of all hurt and killing, "for the earth shall be full of the knowledge of the Lord as the waters cover the sea"?

On another day no one saw Susie fall into the swimming pool. She could not find a way out and very nearly succumbed. At the last moment I heard the feeble paddling. Just the tip of her nose still broke the surface. As I lifted her out she fell exhausted on the earth. I then carried her home to sleep off her tiredness, her beautiful brown eyes speaking her gratitude more than tongue could ever tell.

Susie and Nate were made for each other. When he cultivated his garden with the little tractor, there one could see Susie held somehow in his lap. When he came home from school, they would renew their sundered ties with delightful displays of affection. When he did his chores, Susie was by his side to help. When studies were to be done, Susie was on his lap, or beside him on the couch or at the table. At his feet, on his lap, by his side, she had all she could ever want. They were all a boy and his dog could be.

How does it happen that we find enclosed in one tight little pup-shaped skin all the secrets of the universe so beautifully revealed, yet not at all comprehended?

One shining fall day the children climbed aboard the school bus to be gone for the day. To while away the time Meg and Susie set out to make the most of the beauty all about them. Off bounced Susie, sniffing, circling, nipping at Meg's ears. What promise the excursion held for both of them. Then a sudden darkness fell. The world collapsed in an instant about Susie's small body as a terrible pain tore through her vitals. Just a little puff of sound it was, a small, sharp snap, gone almost before it began. The leaden bullet, accidentally let loose by the gun carrier, propelled mightily by the force of the exploding charge, did its deadly work without a twinge of remorse. It was made for this job.

There was nothing anyone could do. Susie could not hold on to the life she had lived so briefly, so richly. She will never be forgotten. Nate placed a crude white-birch cross above her grave in the white-oak grove. He and the others were brave and staunch. Yet at night no small warm body is wriggling under the covers at his feet. In the morning no little quick tongue licks him joyously awake. It is hard for everyone to understand why so soon a boy and his dog have to be separated forever.

GOVERNMENT STANDARDS ❧ FEBRUARY 1985

Recently, away from home, I stopped at a highway restaurant for a meal. At an opposite table sat a family, each member with a steak dinner, including generously buttered baked potatoes. I could not help observing one of the party return from the bread table with a dish piled high with butter, which he spread in gobs all over his already buttered potato, and he ended the ritual by covering his steak as well with a thick layer of it. By then food had lost its appeal to me. I felt myself in the presence of a human being who was taking his own life by degrees, digging his own grave with his teeth.

This scene reminded me of an acquaintance who, in the middle period of life, was brought low by a severe heart attack. Nature told him that he was not living right. He did not smoke, but was quite a bit overweight, led a sedentary life, and knew practically nothing of nutrition. During many anxious moments, life and death nipped and tucked. His lease on life was finally graciously extended, and he is still with us.

For the first time in many months, I saw him the other day at his place of business. He looks like a new man. He is slim and trim, upstanding, vigorous. He walks several miles every day, come rain or shine. He now knows a lot about what he should and should not be eating. Aside from his damaged heart, he seems the man he should have been from the start.

How often in our world of today do false encouragement and temptation and our ability to live on the fat of the land bring us down. One is reminded of Ecclesiastes 6:7: "All the toil of man is for his mouth, yet his appetite is not satisfied."

My reborn acquaintance, now more knowledgeable about nutrition, has a new appreciation for Walnut Acres. "Do you know," he asked, "that you folks were forty years ahead of your time?" Of course a comment like that can be music to one's ears. Being what I

❖ PRELUDE ❖

Dissatisfied with a conventional life as a college math instructor, I travel to India seeking inspiration and discover my desire to live close to nature.

Left:
Not long after our marriage, Betty and I both teach at Woodstock School near Mussoorie, India, in the foothills of the Himalayas. It is the custom to wear Indian attire for special occasions.

Below:
In 1946, eager to begin a new life, we buy the farm called Walnut Acres. Daddy Morgan, Betty's father, comes to us after forty-five years of mission work in India. (Left to right: Marjorie Ann, Paul, Daddy Morgan, Ruth Carol, Betty, and Lassie)

❄ WINTER ❄

The black walnut trees that abound on our farm were thought by the early settlers to be signs of a good limestone soil.

At first we heat and cook with wood only. I saw fallen tree branches by hand, then use the circular saw on the back of the tractor to cut the wood into usable pieces.

In the winter we can look out of our house at the one-hundred-year-old farm lane, too narrow for modern traffic, and the snow-covered stream that meanders by its side.

Curbside service: Pull right up to the entrance of our very first store, next to this Amish buggy. Only natural foods are produced and sold here.

After a hard winter ice piles up on the banks of Penns Creek. (Left to right: Jocelyn and Ruth Carol)

Ralph the duck falls in love with Tinker the dog, but Tinker holds only disdain for this untoward development.

In winter the snow cover is deceiving. Multitudes of living things quietly prepare to rise again.

🌿 SPRING 🌿

Above:
Sharing beats shearing:
Some lambs are rejected by
ewes and must be bottle-fed.
The little shepherd here is
Ruth Carol, our second
daughter.

Below:
The shepherd shears his
woolly friend in spring.
We send the fleeces to a
blanket mill, where they
are woven into soft,
beautiful blankets.

Above left: With our first tractor I prepare to work the fields for our first spring crops. *Above right:* A horse appreciates a whole ear of dried corn, especially when offered by a tiny child (Ruth Carol). *Below:* For years we do all our farm work with two teams of horses. Here the big, gentle Belgians, Mollie and Prince, are directed by our first daughter, Marjorie Ann, at age four or five.

In early spring a flowering fruit tree stands in a field of grain that was planted during the preceding fall.

On a fine spring day during the early years, I gaze with a strong feeling of kinship over the rich, rolling fields. Penns Creek lies in the background.

❧ SUMMER ❧

Top: Helpers George Richard and Richard Nellis haul early crops of peas to the viner by wagon. Pitchforking vines, pods, and all is an arduous job. *Bottom:* Peas and vines are fed into the machine hopper and drawn into the viner. Revolving paddles inside a large drum beat the pods until they open. Peas fall into containers; spent vines fall off the far end and will become compost. We try to return to the soil everything usable that springs from the soil.

Our helper Bill Newby prepares to cultivate rows of corn with Mollie and Prince and the old cultivator.

Cultivating carrots and beets in stony soil is difficult. Here Ab Bojarsky leads a horse between the narrow rows while I manipulate the cultivator.

Above left: Our third daughter, Jocelyn, ponders life's glories. *Above right:* Our bees, whose attention I divert here with a bellows-type bee-smoker, are never fed sugar. Enough honey is left in the comb to sustain the hive through the winter. *Below:* The old log springhouse, once home for an early settler's family, serves as the family refrigerator for many a year and finally becomes my hideaway.

At times the road leading to Walnut Acres disappears beneath swollen waters. Because of too little advance warning, the foodstuffs stored on the lower floor of one of our barns are ruined this day under four feet of surging water.

Horses return from the field and take in huge draughts from the old wooden water trough; sheep huddle in the hog-pen shade; chickens peck endlessly, living a natural life close to the soil—all happens under the observing eye of Lenore Keene, our niece.

❧ FALL ❧

For many years we make our own apple butter outdoors in a great cauldron, boiling down the cider and apples in the midst of exquisite scents. (Left to right: Paul, Ruth Carol, Marjorie Ann, and our friend Kit Haines)

In fall the field corn is ready for harvesting. At first we do it all by hand, one ear at a time. Then we get our first corn picker. Tractor drawn, the picker pulls each ear from the stalk and husks it. The elevator carries the husked ears above, then drops them into the truck bed.

I maneuver our McCormick-Deering reaper-binder as it cuts and ties the sheaves of ripened grain.

Fodder for the cattle is hauled from the field along with Marjorie Ann (left) and Ruth Carol, who like to hitch a ride.

In our early days we harvest all vegetables either by hand or with the aid of a simple potato-digging device.

The family enjoys a Sunday afternoon wagon ride around the farm with the cousins. (Left to right: Daddy Morgan, niece Winnie Keene, sister-in-law Elsa Keene, Betty, Ruth Carol, Marjorie Ann, Paul, nephew Jim Keene, and niece Lenore Keene)

am, I pretended to demur, however feebly and unconvincingly. But as I returned home my mind was playing over the long-ago years. For in a small way I had probably been helpful in awakening the government establishment in Washington to new ideas about nutrition.

The Food and Drug Administration (FDA) has most important work to do in the food supply. It has tons upon tons of regulations covering all sorts of food products. For these products standards are spelled out, to which commercially prepared foods in interstate commerce must adhere. If the standard for peanut butter states that blanched peanuts must be used in making peanut butter, then the more wholesome unblanched peanuts may not be used if one wants to call the product peanut butter. In aiming for a minimum below which manufacturers could not fall, too often the standard at the same time, in a sense, set a maximum above which the manufacturer could not rise.

It was here that we first came head to head with the FDA. In making peanut butter we used only whole, unblanched peanuts. The important vitamin-containing red skins and peanut heart (or "germ") became a part of the product. We used seed peanuts that would grow again, not seconds, splits, or nuts unsuitable for sale in any other form.

The standard allowed use of from 10 to 15 percent of nonpeanut products like sugar, salt, and hydrogenated fats. We added to the whole unblanched peanuts only a trace of salt if desired.

Upon learning how we made our peanut butter, a zealous FDA informed us that we would be required to label our product imitation peanut butter. We asked our thousands of customers if they would express themselves on the subject to their representatives and to the FDA. Many letters went to Washington, the FDA began to see the foolishness of its peanut-butter standards, and because of the uproar these standards were in time changed to allow whole peanuts, red skins and all, to be used to make peanut butter.

This was one of our earliest forays into government circles, and it established our name in an unusual way. Our country-wide interested "market" became something to be considered when food standards were being set, for a new way of looking at food and nutrition was on the horizon, one whose essence was common sense.

Then we began to make spaghetti, macaroni, and noodles with whole-wheat flour, supplemented with other natural nutritional substances. These products did not meet the current FDA standards, and so we actually labeled them artificial products, and we sold tons of them.

Again, ultimately, government standards were changed, to permit manufacture of far more wholesome whole-grain pastas, with nutritional additives. Before this time pasta manufacturers were forced to make these products from nutritionally inferior white, "refined" flours. This whole matter became quite a laughingstock among persons acquainted with it. Again the new age spoke and the old passed away gracefully. No more will we be required to label more nutritious macaroni "artificial."

We were, I think, the first, and for a time the only ones to claim that the beef used in soups and stews was from "organically raised" animals. How often we were required to make trips to Washington for label approval. We were asked to furnish an affidavit for each animal used in making these products, explaining how it was raised naturally, without the synthetic props so common in our day. Even at that early date we would find copies of *Organic Farming and Gardening* magazine on the desks of the USDA people we visited in our work. In their hearts they knew that we were a part of the wave of the future.

Walnut Acres seems to have borne the brunt of the battle over food labeling for small producers in general. Every few years proposals would come up in the Senate that, if adopted, would have put us out of business. Proposed bills would have required us to analyze every batch of every product we made in order to determine the percentages of major nutrients therein. Doing so would have cost us about a half million additional dollars at a time when our annual profit was just a few thousand dollars. We have always insisted on telling what we know about the way in which our products are raised, about exactly what is contained in each one. At several Senate hearings we were the only ones to raise our voices about the absurdity of situations in which consumers knew exactly how much sodium or protein a product contained but knew nothing about the way in which the product was raised and prepared and what poisonous or suspect additives were used in production. We pled for broadening this aspect of the labeling requirements. We pointed out how absurd and misleading it was to be able to mix shoe leather and glue and make it come out with a dandy nutritional analysis.

All this controversy has a funny side. Several years ago I was called to testify before a Senate subcommittee on nutrition, chaired by Senator George McGovern. I found myself on a panel made up primarily of representatives from nationwide organizations like the National Cattlemen's Association, the National Broiler Council, and the National Food Processors Association. There I sat, representing little Walnut Acres in this august company, espousing a point of view somewhat different from that of the others. I handed out copies of my statement and of the Walnut Acres story, which was later published in the official report of the hearings.

Some time later I received my copy of this two-hundred page document. Every word of my statement was there, including a copy of the Walnut Acres story. There was a photograph, taken over thirty years ago, showing our second daughter, Ruth. A cute toddler in overalls, she was shown sharing her milk bottle with a little orphan lamb who had skipped down to the house, bleating. How his tail waggled in ecstasy with every gulp! The two of them used the same nipple, taking turns. We wondered what future scholars and researchers would make of that bit of Senate testimony, a homely touch in an austere publication!

Recently someone sent me a clipping from *The Wall Street Journal* in which Hodding Carter III, formerly in the Carter administration, in his column "Viewpoints," asked, "Remember When Tomatoes Tasted Like Tomatoes?" He discussed the supermarket tomato, referring to it first as a baseball, then as papier-mâché, finally as a sponge. The supermarket peach, he said, could have been used by David to triumph over Goliath. His expression of sadness for the millions of youngsters and adults who in our advanced, modern, urbanized world will never know what real food tastes like brought tears to my eyes. I wrote to thank him and to invite him to visit Walnut Acres, perhaps in the next tomato season. How many of us have sold our birthright for a mess of plastic pottage?

Nowadays the word "natural" is used universally, albeit loosely on occasion. People everywhere seem to be relearning that which should never have been forgotten. For a time most of us were cajoled into leaving the right, the sensible path, but we are finding our way back again. How often we must return to where we started in order really to move ahead. We must simply refuse to try to build and sustain our lives on rubbish instead of on richness. And we will find ourselves moving from wholeness to wholeness, as once more Earth's sustenance and completeness flow through our beings as they were meant to flow.

MYSTERY OF THE SEED 🌿 JUNE 1982

If I hold in my hand a bag containing a pound of small, light, fluffy, unprepossessing tomato seeds, what do I really have? It is impossible to comprehend all the ramifications of that question, for in that container lies the mystery of the universe itself. You remember Tennyson's poetic comments:

> Flower in the crannied wall,
> I pluck you out of the crannies,
> I hold you here, root and all in my hand,
> Little flower—but if I could understand
> What you are, root and all, and all in all,
> I should know what God and man is.

If the universe can be found in a grain of sand, even more may it be found in a tomato seed.

As soon as planted, and under proper conditions of moisture and temperature, something within the seed tells it to align its forces in the direction of forward-moving growth. It asks no questions but quietly goes about its work of growing, with the end of reproducing its kind. Its drive in this direction is phenomenal, as is its determination to persevere until the very end. From start to finish it demonstrates the power of positive thinking. An illustration from earlier days tells of its resolve.

Many years ago, after completing our work in an organic farm school, we had moved to a rented farm to live until we could find a farm of our own. In that year the crops looked most promising. We had corn and oats and wheat and hay-grasses, plus grapes and peaches and cherries and berries, plus a large vegetable garden kept in decent order. An ornery team of horses provided the only work power aside from our own.

We worked hard, felt with joy the rippling of young muscles, whistled while we worked. Our very precious, very first child, Marjorie Ann, was all of three years of age. She was the one who learned early how it felt to be bowled over and "flapped" by the king of the poultry flock, the old, exulting red rooster! (He never got the chance to do *that* again.)

That was the time when a mature lady from the city, having heard of "these new organic farmers," showed up at the farmhouse one day to converse about this and that. Just then, stripped to the waist, standing in the old wagon behind the horses to which I had given free rein, singing at the top of my voice to drown out the rattle of the wheels over the cobblestones, I came roaring in for lunch. As I pulled up short by the barnyard gate, the lady, obviously taken aback by this unlikely performance, was heard to mutter quietly, as if to herself, "I expected something less virile and more intellectual." Oh well. Someone must do this world's work.

Shortly thereafter, on the Fourth of July of that same year, we watched thunderous clouds overtake first the afternoon sun, then the whole sky. Darker and darker they swirled, increasingly ominous. We battened down all hatches, prepared for the worst. It came. The furies took over, howling demonically. Hail fell with shattering blows for hours on end. And then the sun set in all its glory. On the following day the very same thing happened. And these were the only days, in all these years of farming, when we have seen truly destructive hailstorms.

Everything lay in ruins. Every crop was destroyed. Young fruit trees were killed, the bark skinned right off. Promise had quickly turned into utter desolation. It was too late to replant all but a very few garden vegetables. The tomato stalks, previously lush and burgeoning, were mere stubs of stems and branches. But a marvelous thing happened. Within several days we became aware of little green buds appearing all over the stubby tomato-plant stems. Soon we had regrown plants, the sturdy roots pushing nutrients madly into leaves, flowers, and ultimately a good late-fall crop. Those were the only plants that survived.

Let us tell you the life story of the tomato at Walnut Acres, from seed to fruit. Over the years we have witnessed tomato growth many times, yet it is ever new and fresh to our hearts. We are small, relatively unsophisticated growers, using much hand labor in planting, tending, and harvesting this delightful crop. We cannot relate the tale without referring to the neighborhood small fry, who love to hang around the edges. We think in particular of the neighboring-farm grandchildren. From toddlerhood on, theirs is a privileged position.

They see and learn many basic things. They will never have to go through life without knowing all about the wonders of a year gloriously red with bushels and bushels of the ripest, sweetest, juiciest tomatoes ever seen. To them the whole procedure is simply a part of life, the way things should be. Each year they see and know more than in the previous year. When small, they begin to help in small ways. They work on an equal footing with the adults, secretly bursting with pride as they stretch their small selves to the limits of their lesser abilities. They learn by doing, which is of course the soundest way.

They may see the seed as it arrives from the growers. They may see the tiny plantings bursting through the soil in the greenhouse. They may help remove the seedlings from their cradles in preparation for planting. They may help in keeping the planters supplied with seedlings, in mixing fish-emulsion solution and kelp liquid in the planting water, and in resetting plants that have ended up in the soil with roots in the air. Such things happen more often as the long day wears on. At such times the eyes cross, the hands lose something of their dexterity, the body struggles to keep all its parts in focus. Over a span of seventy years or so, with men, women, and children working together in jolly, uplifting unity, all become a part of one contributing and fulfilled family. One's whole being smiles. Work becomes play. At night, sleep is deep and sweet.

The baby plants are grown for us by a local greenhouse, with heat supplied by a nonnuclear generating plant about forty miles from the farm. When we receive the plastic trays with many long, thin, deep pockets, the plants in them have grown from six to eight inches high. They may be lifted readily from the pockets by merely tugging on the plant tops. The planting material remains entwined in the roots and is held together in a two-inch, upside-down cone of soil that stays in one tight clump. The roots are never bare, and so the plant keeps right on growing in the field with very little setback.

It is a great day when the plants arrive at the farm, tray after tray after tray of them. Only one thing in this world has the odor of tender, young bright green tomato plants. The machinery has been made ready for the year's work—the planting machine runs for only two or three days and then goes back into storage for another year. Seed planting is timed so that the plants are ready to go to the field just after the frost that we hope is the last frost of spring. Tomato plants are so easily killed by low temperatures that every year some early planting farmers lose some of their plants to frost. Those who do must start again—a bitter fate.

Rows are planted three feet apart, with plants perhaps fifteen to eighteen inches apart in the rows. The planting machine is rather a weird mechanism, cleverly conceived. The tongue is hitched to the tractor drawbar. Two furrow-opening metal shoes automatically open two furrows as the contraption moves slowly forward in the field. Two wire mechanisms with rubber-coated grippers open at just the right time, to grasp the next plant placed upon them by the planters. From the top of the revolving plant-wheel, the plant, held tightly, is carried securely down to the open furrows beneath, root down. At the right moment the grippers suddenly release the plant, which drops into the furrow just as a bit of water pours from the tank, soaking roots and soil.

Still farther back are four very hard metal seats, which seem to float along about a foot above the ground. The planters each sit on a seat with unbent knees, holding legs parallel with the earth's surface. Folded burlap bags are generally used as cushions, lending small comfort after several hours of bouncing. Heaps of plants pulled from their plastic birthplaces are placed in boxes before the four planters. The whole apparatus is supported on two large cover-wheels. Two drums of planting water holding fifty gallons each ride on the front of the frame. Each has in it the proper portion of fish and kelp.

Well, here we go. The tractor driver has a field marker to show him where to place the tractor wheels for the return trip, so that each pair of rows will be evenly spaced. Everyone is seated except for the person who will be walking behind the juggernaut to check for errors, for stones, for clogged waterline sinuses. The planters have burlap bags on their laps. They pull a handful of plants from the storage boxes in front of them and load their laps. Ah, the thrill of the annual tomato planting festival!

The tractor starts crawling forward at the slowest possible speed. The ground wheels start to turn, and also the plant-gripping wheels. The furrow openers dig in and start to turn. The planters take turns laying a plant on the grippers as these gadgets revolve in front of them. Suddenly the grippers grab and hold the plant to prevent its falling off. When the plant has been carried from the planter around the wheel to the soil, the grippers let go of it just as the roots are being placed at the bottom of the furrow. Just then, a measure of water squirts into the furrow all around the plant roots. The large cover-wheels are so made that as they pass over the newly dropped plants an opening between the two wheel halves keeps the plants from being crushed while the sides of the wheels themselves push the earth back into the furrows around roots and stems. And so each plant is planted and firmed in the waiting soil.

The whole thing runs smoothly. Something gives a metallic click just when the water is about to squirt into the furrow, the signal that it is time to place another plant on the top of the gripper wheel. You had better be ready, or you will leave a skipper in the row. If thirty thousand plants are to be set that day, each planter handles seventy-five hundred plants in ten hours or so, which averages out to about five seconds per plant, including refilling, reloading, turning around, and so on. Of course, positions are traded about during the day.

How lovely is the sight after a day or two. Let there be a nice rain, and the plants are hardly set back at all. They look as if they had always been there, their roots reaching out into a far wider world than they could have known earlier. Days pass. The plants are cultivated a number of times, sprayed with a water solution of sea kelp to strengthen them. One can almost see them growing. All at once blossoms begin to appear, then green tomatoes follow, and finally one sees the first blush of red. This is the sign that the baskets must be brought out and spread abroad in the field for the pickers.

A pound of tomato seeds, about eight thousand in all, will plant roughly one acre, to produce up to twenty tons—forty thousand pounds per acre of luscious, unchemicalized, untainted tomatoes.

Hand picking is a kind of luxury now. In some areas monstrous, clanking machines go through the tomato field picking the entire crop in one swoop. Tomatoes being what they are, ripening one by one over a long season, the picking machine is bound to gather in a multicolor crop, ranging from rather green to rather red.

Of course, with genetic manipulation and the like, tomatoes can be built that have less of a ripening spread, so that the variation in color will be less than in our fields. Also, by manufacturing tomatoes with a tougher skin to withstand the steel-fingered robot, you are less apt to find all that juice at the bottom of the hauling truck as tons of them are jounced along in one huge lot, without benefit of baskets, to a distant commercial cannery. (Ours is right here.)

Certainly it is easier and less costly to crawl though the fields with an unremonstrative monster, clearing everything at one picking, than it is to arrange for pickers. In our area the Amish folk, who live all about us, do most of the picking. If we did not respect their dislike for photographs, we would show you the fields full of these superb pickers at work. Because they are a horse-and-buggy people, they have no automobiles. We must, accordingly, arrange for their transportation by motor vehicle, with teams of men and women coming from as far away as twenty-five miles. In rainy weather one is never quite sure what to do. Amish people do not have telephones. Problems can arise. If they come to pick and it starts to rain, they will pick anyhow. It is all very messy then; such devotion. (Of course in rainy weather, the big picking machines bog down, conquered by the mud.)

A good picker can pick from one hundred to one hundred fifty baskets in a long day. At fifty cents a basket, that means fifty to seventy-five dollars a day. They would do better in some ways to be professional people, which must be easier on the back. Yet with feet often bare, with long-skirted and panted bodies otherwise fully clothed, dealing in such proximity to Earth and her fruits must do something for the soul that makes the backbending labor supremely worthwhile.

Of course, chemicals can be applied to discourage fruit flies and other pests in large tomato-handling plants, by spraying the fruits with insecticides—methods that, fortunately, we have never used and intend never to use.

Further, marvelous gases of one sort or another can make green tomatoes become as if by magic red tomatoes. In fact, many farmers now pick *all* their tomatoes in the green state, to have them appear later in urban markets supposedly "field ripened" (rather sickly looking reddish blobs of vegetation lacking in taste).

Ask the grandchildren about the technological practices. What know or care they about all sorts of synthetic fertilizers poured onto the soil, about poisonous chemical sprays applied to the plants, about picking other than by hand, about insecticides to be sprayed on the fruits themselves, about strange gases that turn green to red, about the sodium hydroxide (lye) used to eat off the peels, to save labor? All they know is the deep, incredible lusciousness of eating an earth-formed, sun-warmed, nature-ripened tomato in the beauty of wholeness.

Who of us, in holding a tomato seed between thumb and forefinger, can even begin to appreciate the power, the purpose, the drive, the courage, the tenacity, the single-mindedness, the breadth, the knowledge, the ability to survive and carry on in that one speck of dehydrated life? We're told that farmers who use sewage sludge find volunteer tomato plants in their fields. Think of the organizing power that seems to see ahead to the finished product, the end result, sending sinuous, seeking fingers throughout the solid earth in the precise directions required to find the most food and moisture, drawing into the plant against all odds the wherewithal to reach its destined end.

Let the odds be what they may, without bitterness or complaint, doubt or vitiating depression, always looking up, tomatoes move in fine fettle to their appointed end. May one dare to use words like hope, promise, spirit in reference to that tiny seed? What are you, really, you insignificant blob of life?

MOLLIE AND THE HORNETS 🌿 AUGUST 1970

Years ago, when life was young at Walnut Acres, we did not own tractors but farmed with horses. Once I was driving my old McCormick-Deering binder, which cut the wheat stalks, bound them into sheaves, and dropped the sheaves on the ground in piles. I had three skittery horses pulling the binder, and as we crossed a narrow bridge, one outside horse was pushed off the bridge and into the stream below. I was a very shaken young farmer, but I knew the horse would be shaken, too. Jumping down to the horse, I uttered a rapid succession of sounds designed to be soothing to beasts of burden. Horses are funny creatures. What a few pretty whispered nothings can do for them! Had the horses chosen to panic, this situation could have ended as a tragedy, with a tangled mass of flesh and steel and twine. But it simmered down into the sweetest little tête-à-tête you ever saw.

We still find in the fields shoes dropped by Mollie and Prince, the mighty Belgian team, our favorites. Mollie had a cute trick. She would hang back just a trifle to let poor dumb Prince do the lion's share of the work. When scolded, she stepped up for the moment, only to slip back soon to her one-step-behind place.

We had to think of the horses particularly this year. It has been a bad year for bumblebees, hornets, and yellowjackets, particularly the kind that build their nests in the ground. Quite a few stings have been felt among those who work in the fields.

More than twenty years ago we were plowing the field just below the barn, with Mollie and Prince in the harness. It was fall, laziness was rampant, and no one felt much like pushing. Old Lassie trotted behind in the furrow, sniffing at the freshly turned earth. It was as bucolic a scene as one could wish for.

Suddenly Mollie stepped out with a great lurch, dragging a sleepy Prince along by the singletree. The plow stiffened itself and began almost to ring with the friction. The plowshare was new and sharp, and so the earth rolled over in a continuous foot-wide ribbon at a supersonic pace.

When things quieted down a bit we stopped to rest. On going back in the furrow to the place where Mollie's interest had become so abruptly activated, we found the reason. We had plowed up a hornet nest, lock, stock, and barrel. The hornets were everywhere protesting this desecration. I don't think Mollie had been stung, but she certainly knew better than to loiter when she sensed the hornets.

The next ten laps or so around were fun for everyone when we came near the uprooted nest. All three of us did a lot of fancy skittering. Our feet were off the ground as we sped by the demolished abode. The field was plowed in record time that day. Thereafter I often thought of holding a cage full of angry hornets on the end of a stick in Mollie's ear whenever she took too much advantage of her privileged position. She had moved beyond the contemporary feminist movement: she expected equal pay for unequal work.

A FARMER'S CREED 🌿 JULY 1978

Birds sing everywhere, from very early morning until the sun retires: robins and wrens and cardinals, orioles and catbirds and other assorted small feathered persons. But oh those purple martins. I hope you too have some of them flying around your house. An unbelievable liquid gurgling fills the air in the vicinity of their nest. It is one of the most uplifting sounds I know. Perhaps some of the next world's music will be spiritually akin to these glorious chirps of praise.

More than three hundred years ago the Moghul emperor Shah Jehan is reported to have said of the Vale of Kashmir, "If ever there is a paradise on earth, it is here, it is here, it is here." He should have seen Walnut Acres on a rain-soaked, intensely greening, sunny late-May morning.

But we are not a paradise of endless bliss, of utter security, of untroubled spirit. Each day, each hour the challenge is new. We constantly probe at the edges of the universe with each new day, to learn how best to match needs to circumstances. The low fields are too wet? Go to the hills, where underground waters subside sooner. None of the plowed fields can be worked because of recent rains? Spread manure on the sodden ground. The carrot fields had beating rains upon them, making it almost impossible for the young plants to push through? Change plans at once; replant in another location, before it is too late.

Change plans on the spur of the moment. Make up your mind immediately to meet unexpected conditions. Shift the emphasis of the whole enterprise for a day or a week or a year. Weigh, balance, adjust, accommodate. Make mental and physical notes every day on how you will do this operation better next year. Keep a thousand eyes open to evaluate and decide on the way you will meet each new situation, each new day. Be a computer with a sense of humor, a philosophical spirit, feeding yourself unconscious data, settling finally on the readout that seems best at the moment when needed.

No time for inflexibility, for the aging spirit, for quibbling with fate. When the cow has died, one goes on from there. Instead of cursing the flat tire, one hustles to repair it at once. Even on the days when everything goes wrong, one comes up smiling at the supper table.

For all this exercise of the spirit is the glory of farming. It stretches one almost all the time to the utmost. It pushes and pummels and molds. Body and mind and spirit remain lean and strong and stalwart. Life becomes a constant battle of wits, a rolling with the punches, a staying on top of circumstances.

The attempt is not to get ahead of Nature; it is rather to keep in tune with her fickle hourly vagaries, dancing when she plays a joyous tune, somber when she weeps. For underneath, one feels the tremendous strength of an unshakable security and promise, of which the hourly changes are but the minutest surface movements. One discovers in farming more about what life is really like than in any other occupation. We're at the roots, the source.

Out of one's touch with the soil grows an all-pervading warmth that rises all about one, enwrapping, heartening, and ennobling, speaking softly to one's spirit words of calm and hope. And so one comes to match this day not only with yesterday, but rather with last year and the year before and with an endless chain of years, until in the light of a lifetime the infinitesimal bumps of seeming adversity are as nothing. That which today appears as playing a losing game is in the end, in the face of eternity, nothing less than ultimate victory. We are blessed with the meeting of our deepest need, the certain knowledge that the universe is good. What better can this life offer? If ever there were a paradise on Earth, it is here.

FINDING WATER ⚘ NOVEMBER 1985

I am not up on my dowsing-rod terminology—witch hazel, forked peach, willow branches, and the like. If and when rare thoughts about such matters come to me from a long past, in memory I see a man who visited our farm from time to time, wearing a broad-brimmed Texas cowboy hat. This man could expound at length on his regular attendance at conventions of a national dowsers' society.

He told us of some who had moved beyond the local scene in their water-finding propensities. They could, he said, study a photograph of a plot of ground many miles from home, and, using their special abilities, tell where water was to be found in the area pictured, the depth at which it would appear, and its volume. It seemed a kind of mail-order water-locating service, whose validity we could never deny or affirm. Our friend was a most enthusiastic and convinced member of this society.

We are told that some have such a way with these forked sticks, such power working through them, that once they locate water as they hold on to the branch arms with each hand, the third arm dives toward the earth with great force. It is claimed that these people simply cannot restrain the branch from twisting in their hands no matter how tight their grip.

One of my brothers knew a person with unusual dowsing ability. While this person looked elsewhere, one could toss a marked coin into a field of tall grass or weeds, and if the general area was pointed out to him, he would walk slowly over and around the indicated space, forked stick at the ready in front of him, and suddenly the free end of the stick would point toward the ground. Sure enough, there lay the coin, hidden under all the greenery.

I've had a try or two at seeing what I could do with a three-pointed, forked willow or peach branch. My efforts at finding water have had results desultory at best. Several times with forked branch in hand I thought I felt a feeble tug here and there. The upheld pointing end did twist down a bit, but certainly not with force sufficient to remove the skin from my fingers. I tried it over a well top. There it appeared to be particularly drawn down. We have so much water flowing from the ground about us in springs, however, that we are really not interested in finding more.

But one extrasensory perception, if that it be, took hold of me many years ago. No witchcraft, no sorcery, no demons, no devil worship; just simple, unexplained, fascinating fun it has turned out to be, with helpful overtones. The universe and I have a special thing going between us. You are about to be let in on a secret in the hope that you too may become part of the mystery, feel its working within you. I don't really want to know how my ESP works. I use it when needed, forget it otherwise. But I do want to share it for what it may be worth to you.

Over the years I have watched farmers, plumbers, and bulldozer operators produce a small pair of stiff wires and with these in their hands tramp over the ground where they are working. They move sedately, eyes straight ahead, apparently in deep, concentrated thought. They hold these wires pointing straight ahead. Something happens all of a sudden: they seem to find something for which they are looking. They retrace their steps, move to another location in the general area, and try once more. Finally, having learned what they wanted to know, they go on with their work. Here are practical people using wires to help them locate the general vicinity of underground water pipes or sewer lines. It must work for them. They would not continue to engage in a worthless exercise.

Many years ago I first tried to locate underground pipes (or streams?) with wires. At once it seemed to work for me. Since then I've used my wires many times. True, at times I am not quite sure of the response. Sometimes I think my imagination works overtime. But overall I have had much good sport with them, and we at Walnut Acres have been helped much by using them.

If the wires really do work, they appear to respond to the presence of pipes of metal or plastic or clay. They even respond, it appears, to unused pipes still lying under the ground. Every ten years or so we must have a septic tank emptied. I locate its pipes and general location with my wires. From the main floor of a house one can also find pipes that run through the ceiling of the cellar below. Many people believe that underground streams attract the wires, too.

I once plotted a dozen underground streams in a field near our house, all coming together at a point directly in the bed of a small stream flowing by the house. It happens that precisely where these underground streams seemed to come together, the small stream itself disappears into the ground in really dry weather. I've poked about with a crowbar there in the stream and, in my amateurish way, found that the bar very nearly dropped into a place like a void. I thought I could hear water rattling underground where the stream made its final entrance into the earth.

Strangely, farther out in this same field in a winter long gone appeared a sinkhole into which one could have set a large wagon. It gave us a scary feeling to ride on a tractor in that area. We finally filled up the sinkhole, and it has stayed that way ever since. We live in an area of limestone caves, and one of these must be down there somewhere. With it all, I generally feel a tiny shadow of a doubt within. I am prepared to admit that I may be fooling myself. If so, I love it.

I'll explain exactly how the pipe-finding is done. I hope that, if you try, it will provide family amusement, with some real successes in addition. I must tell you, though, that some people seem able to get a response and others get no indication of anything unusual. Only 10 to 20 percent of those who try seem to have positive results. Don't let those who can do it look down their noses at those who can't.

Take two regular, thin-wire coathangers with no cardboard sections. The bottom wires are about fifteen inches long. With pliers or tin-snips, cut through each bottom wire at one of its ends, just before it begins to turn up. Now return to the uncut ends of the bottom wires. Measure about five inches from the lower ends of the wires leading to the top. Cut off there and straighten the wires into an L shape. Both wires should be cut to the same length. You will now have two L-shaped wires about five inches by fifteen inches. Be sure you get the kinks out. The L must form exactly a right angle, to allow the wires to swing freely, when and if they do move.

You will be walking along slowly, holding the shorter ends of these wires in very limp fists. The wires must be very loose in your hands, able to swing about freely as called upon. They must also be held so that the long top wires, when pointed out in front of you, do not rest in any way on your index fingers. The top wire of the 90-degree angle should be about a half-inch above your fists.

The most important rule is to hold the wires with the long side parallel to the ground at all times; most who try are not careful enough to do this. If the ends of the wires that stick straight out in front of you droop down, and then the wires do start to spread apart when you come to an underground pipe, the weight of the hanging-down wires will help prevent the necessary spreading. Nothing mystical as yet; just plain physics. The two wires will be parallel to each other and at the same time parallel to the ground.

Here's the way to begin. Seek a place where you know a pipe goes under the ground or floor. Go back six feet or so, then face the pipe so that you will be crossing it at right angles. Move very slowly forward, with a mincing step. As you do, your upheld wires should mysteriously exert a tugging at the far ends. As you draw near the area of the pipe, the wires should begin to spread apart. When you are standing directly above the pipe, the wires should form a straight line above and parallel to the pipe in the ground. They will sway about a bit, but they will have been moved by an unseen hand. As you go beyond the pipe, the wires will tend to return to the position from which they came.

With wires in hands and arms by your side, stand straight and tall. Let your upper arms hang straight down. Then raise your fore-arms until they are at right angles to the upper arms at the elbows. Bring your wire-holding fists to within six inches of each other, long wire sections parallel to each other and to the ground. Looking like a Martian with misplaced antennae, you are now ready. Try it first when the neighbors aren't looking.

Be absolutely certain that your wires are above your fist tops and are held very loosely so that they can swing. Mince gingerly for-ward, taking short, slow steps. Time will tell if it is or is not going to work for you. If it does, you will thrill to the feel of that special tug-ging at the ends of your wires. You will have just touched another world.

Our interest in this phenomenon really came into focus years ago when we wanted to build an addition to our cannery. We needed more water. We hired a hydrogeologist to tell us where to drill to obtain a hundred gallons of water per minute. He studied maps and the lay of the land. He looked for all the signs his trade suggested as indicative. Then he drove a stake at the spot he chose as most prom-ising. The driller came in, set up, and drilled one hundred seventy-five feet down in the rock. There he struck water, but with a flow of only ten gallons a minute. As he went still deeper, his drill bit be-came stuck in the rock. Nothing he could do would break it loose. Finally he asked if I would tell him where to drill, because he had to start all over again.

I knew that some users believed the wires could help locate un-derground streams, and so I brought mine out to the site. First I thought I'd locate something that made the wires separate. I marked this possible stream in several places as I followed it by crossing back and forth over it at ten-foot intervals. Then I found another stream (at a different level?) at right angles to the first. Where these two underground streams appeared to cross, I asked the well driller to drill once more. This placement was almost too much for him, because the spot I chose was only about fifteen feet from the first hole. But he drilled there.

At about one hundred fifty feet he was getting ten times as much water as he had earlier at one hundred seventy-five feet. The flow was at last a hundred gallons per minute. We've used millions of gallons of water since that time, and the well has never even come close to running dry. Perhaps it comes underground from the mountain on the horizon less than a mile away. There are no polluting industries for many miles around. Our fields have never had poisonous, polluting chemicals used on them. Departments of health require monthly testing of all our water, and it has never failed to pass easily.

One can argue that the whole thing came about just by chance. Some believe that the universe just happened, without plan or purpose, that it was formed by chance. I cannot accept this assessment. I like this quotation from Albert Einstein: "The most beautiful and most profound emotion we can experience is the sensation of the mystical. It is the sower of all true science. He to whom this emotion is a stranger, who can no longer wonder and stand rapt in awe, is as good as dead. To know that what is impenetrable to us really exists, manifesting itself as the highest wisdom and the most radiant beauty our dull faculties can comprehend only in the most primitive form—this knowledge, this feeling is at the center of true religiousness."

TWO REMARKABLE MEN 🌿 JANUARY 1984

The picture, the calendar, the mirror remind us about how constant change is within the framework of relative stability. The seasons have once more changed. The new year is a slate on which each and all of us will write both our hopes and our fears, our trust or our mistrust in the goodness of life.

Once I talked with a young man who had invented a new kind of mirror in which we see ourselves as we really are, and not as a reverse image, left to right. In his mirror, named the *rorrim,* one sees one's left eye as being on the left side of the creature in the glass, not on the right side. We feel it is our everyday mirror that should be called the rorrim, and vice versa. But no matter in which of these glasses we see ourselves, the impression is probably much the same. The marvelous thing about growing older is the escape from the tyranny of the mirror. The wrinkles and blemishes are there, but we really don't see them. Rather than looking at our faces, we now appreciate more the feeling of life surging through us, and are thankful. There is too much to be done, too much that is wondrous to be experienced, to be held back by a mere piece of glass.

Yet we know that, like the people who have touched our lives and gone, our mirror image will one day fade. Two men, whose influence on me and on the lives of many others was remarkable, have recently died. When I was young, these men already seemed old to me. I sought them out and sat at their feet, literally and figuratively. With but minor variations they and I seemed to ourselves to be in tune with one another. To me they exemplified the full life, growing out of wholesome respect for and adherence to the laws of Nature. They seemed to want to live simply, carefully, and wholesomely, so that worries about bodily ills and distempers could be at a minimum. Thus they were able to spend their copious energies in showing humanity a better way to live.

It was June when I last heard from my dear friend Milton Wend, who said he had just returned from his seventieth college reunion, as the only survivor of the class of 1913. It gave him a "peculiar feeling" to realize that at almost ninety-two he was the oldest of the several thousand alumni present. He wished that he and I didn't live so far apart.

The shade of Milton looks over my shoulder as I write, because this old chicken-house office is where Walnut Acres got its start, at his instigation. Thirty-five short years ago we visited him in New Hampshire. We had been at Walnut Acres just two years, and we needed direction and income. He had written *How to Live in the*

Country without Farming, which offered some hope. Milton Wend, we found, knew much about many things. He had his own stone mill, and he ground flour for the neighbors. He thought we should try doing that, too. As a result, this same chicken house became our first mill. Yes, he is with me now.

In July a letter came informing us that Milton had died quietly in his sleep, with neither pain nor indignities. He was reading, thinking, making notes until a week before his passing. He had never spent a night in a hospital, never had medical bills except for small emergencies, never claimed any medical insurance. He didn't even use glasses except in the last year or so, having done eye exercises successfully.

In August we learned of Scott Nearing's passing at age one hundred. We've read that thousands of persons in our country are this old or older. Perhaps not all these folks live so well to the end as did Scott Nearing. He was never ill. Helen Nearing said that Scott's body just wore out. He had no complaints, no pain. He just stopped eating and drinking. He drifted off easily, like a leaf from a tree.

Long ago, Scott Nearing had left university teaching and city living to homestead in New England. In the spirit of self-sufficiency the Nearings grew their own food, cut their own wood, built their own roads and houses. Their gardens, their sweet peas were out of this world. Their book, *Living the Good Life,* reached thousands who sought a better way than city living. When we visited the Nearings some years back, Scott had just finished digging a hole for a large farm pond, using only shovel and wheelbarrow. When he was in his nineties, he and Helen together built a new house. They lived frugally and simply. *Newsweek* wrote of them that they were early leaders of the back-to-nature movement.

On his last birthday, the neighborhood children gave Scott a carrot cake with one candle on it, and a hand-painted banner with the words: "The world is a better place for 100 years of Scott Nearing."

With the demise last year of these two men, the world is a poorer place. Yet this is the way of life. We can but be thankful for their having risen out of the sod for their speck of time. If we find value in their work, it is ours to help carry it on, as they become a part of the bank of history.

SEEDTIME AND HARVEST 🌿 NOVEMBER 1958

I wish you could see our cornfields this year. We got seed from a good organic farmer friend. It is open-pollinated corn and considerably higher in protein than the newer, hybrid varieties. We were unprepared for the growth this corn would make. By the mill-house door you will find transplanted from the field two stalks that were twelve feet five inches tall in the field. A six-foot man cannot reach the bottom of the higher ear of corn.

The cornfield itself is a forest. No one in the area has ever seen such growth. We are as pleased as if we had done the growing instead of just the planting and tending. It is a tall variety, obviously, but the tremendous growth would be impossible without a reservoir of fertility down under. Next year we will not plant the corn rows as close as they are now. This corn should have more room. As farmers we will hope for no early frost. Farming seems to be always a race to squeeze between two limits. In a way it is fun—and in a way it is serious. The farmer must always be prepared to have some poor crops along with the good ones.

The soybeans are a good example. This year the field can be referred to either as soybeans with weeds or weeds with soybeans. The cool, moist growing season boosted the weeds and discouraged the beans at the start. The beans will not be harmed in any way except in quantity. It cheers us in looking around to notice that we are not alone in this condition. Some years are worse than others.

The fields are almost ready again for planting the wheat and rye. They have had the initial preparation. By the end of September these crops should be in the ground. Once more we will have come full circle. As long as the Earth shall stand, seedtime and harvest— seedtime and harvest—the very words are a part of us. What grand assurance farming provides.

The reflective time of year is at hand. In spring one dreams brightly, glowingly. In fall one dreams pensively, wonderingly.

Spring looks forward to pleasant outdoor activities. It is a time of growing, expanding, of moving out. Fall is a time of slowing up, of moving in. The autumn haze envelops the fields, the year draws itself in, life all about begins to seek its roots. We find ourselves with more time to seek our own roots and to meditate on that which is to us all mystery: life itself.

It is good that things are so ordered. Constructed as we are, we would not be happy if we knew everything. If the creature were as great as the Creator we would see no need to look up. Mystery is the essence of all things. Without mystery, no wonder. Without wonder, no growth. Without growth we are no longer human. In a sense, then, we feed upon mystery and it becomes our life blood. Our hearts beat to a mysterious rhythm. With all our knowledge we know but in small part. We have not yet reached the time when we can know even as we are known.

But we would not want to be made in any other way. This moment tries our strength. This trial tests our faith. This challenge gives us the chance to expand to the limits set by our minds and our hearts.

I have the feeling that if we knew how to live to the full with ourselves and with others, then life, even though filled with mystery, would seem complete. We would know how to store up life's essentials so that we could ever die back to our roots and ever spring forth again with newness and freshness and vigor. The glory, the wonder of the fall season would set up within us sympathetic vibrations such as we rarely feel in our too often dreary, dismal existences.

We have not reached up enough for life's basic ingredients: truth, beauty, and love. We have thought to build with unsturdy blocks. And we have found that our roots can be no larger than our stem and leaves provide for. We are what we eat, even mentally and spiritually. We have little need for spiritual reducing diets. Perhaps this fall will find us, then, reaching up so that we may reach down. Perhaps with our hearts we will learn to hear the message whispered through the aisles of the cornfield as we see the leaves tremble to its passing. May our days reflect the depth of the autumn sky. As we live amid the incomparable glory of the golden hills, may we find no room for dullness in our souls.

PITY THE POOR HEN ❦ JULY 1958

Several weeks ago a chicken farmer called to find out if we wanted some chicken manure for our fields. Having heard about his venture, I decided to pay him a visit, to see things at first hand.

The new, long block-and-aluminum building looked quite attractive from the outside. Inside—well, it is difficult to describe the ambiance. The atmosphere was a combination of hothouse-library-reading room-hospital. A sign on the door said entrance was strictly forbidden—disease germs on the shoes, you know—but someone had told me to walk in despite the warning, and so I did. No noise, no quick movements—the hens must not be frightened.

Two long, spotless concrete paths stretched away into the distance. At the far end of one of them a man was silently pushing along a rubber-tired egg cart. (This was the final collection of the day: the hens would now be tucked into their wire beds for the night.) In a room at the left a teen-age girl sat weighing and sorting mounds of white eggs in an aura of muted popular radio music. The main sight was the hanging cages of hens. It was all somehow so ethereal, so gossamer, made more so by the long rows of white, angelic hens in their dainty, light, airy cages, each in her narrow cell forever laid—and laying (apologies to Thomas Gray).

Let me draw you a picture of the cages. Made entirely of stiff wire, they were suspended in midair about four feet above the floor, a common partition dividing each pair. The four long rows of cages stretched, one on each side, along the two interminable walks. Several thousand cages each held one white angel. In front of these wire-mesh cages ran endless automatic feed troughs; in back ran water troughs, with fresh water always flowing. Droppings fell through the wire-mesh bottoms into a pit. A slanting wire rack under the cages allowed the droppings to fall through but caught the eggs, which then rolled forward into another trough. Unfortunately, the eggs had to be touched by human hands to be placed in the collection basket—a shortcoming soon to be corrected, we presume. Let no one think that precious space was wasted. Although by great effort the hens could turn around in their cages to go from feed to water, from water to feed, it was not without sacrificing a few tail feathers that got rubbed off on the cage side from time to time. They had room enough to stand reasonably straight. Perhaps a crude, old-fashioned chicken stretch-and-flap would not be much appreciated by these modern, sophisticated, urbane biddies even if it were possible. If the open-mesh wire bottoms made the hens teeter a bit now and then, they were blessed by not being able to fall far in either direction. Of course, they were saved the trouble of having to grasp onto anything as outmoded as a roost or a tree branch.

No longer needed by these hens are the things that mean so much to fowls in the natural state: live green grasses and plants in great variety; fresh insects and bugs and worms, alive until just the moment before being plucked. Missing are the other ground-given goodies that only hens can appreciate, to say nothing of the running and chasing and flapping and dusting themselves deliciously that is a part of a full hen life. And how about the mothering of cuddly chicks? The assorted missing small things may help more than we know to make a difference between complete and vital eggs and a glued-together mess of tasteless eggy substance that somehow is wrapped into a package of the same shape, yet must by Nature's rigorous standards be unbalanced.

Instead, the hens have only air furnished by fans that run all the time at set periods, mechanically supplied running water, scientifically compounded feed, and a quiet, shady, restful, cultured atmosphere. Not a worry in the world now, not even old age.

Fluttery little transmuters of feed and water into eggs and manure, they never heard that they are what they eat. You can't get blood from a turnip. One year to live: then soup. Of course, if illness happens to attack in spite of the super-pure conditions, no one would know much about it until our white angel drooped a little. It is nice to know that she will not droop long, for efficiency calls for full cages of up-and-coming, lay-or-bust, now-or-never, do-or-die chickens.

Oh yes, the eggs. They go only and entirely to a very well-known and popular chain of highway restaurants. These restaurants will not buy eggs produced in any other way. Just thought you'd like to know. From the quiet, restful atmosphere of the egg factory to the quiet, restful atmosphere of the restaurant they go, fully synthesized into shining beauty.

I heard later that the poor chicken man cashed an egg check for seven hundred dollars, put the money in his wallet, jumped on his tractor, went out to plow a large field, and returned to find that he has lost his wallet; he had plowed it under. To hunt for it must have been like looking for an ounce of naturalness in a basket of these pure-white triumphs of the chemist's art.

Speaking of unnaturalness in producing eggs, let's look at the way in which other chickens, those bred for meat, are fed in their entire lifetime of thirteen weeks or so. Never do they feel the ground under their feet, never do they see a blade of grass. In cavernous buildings they spend their brief span on earth roaming under artificial lights and eating at common feed troughs with their brothers and sisters. And what do they eat? This is a list of the contents of a commercial broiler feed as found on a feed-bag tag: chlortetracycline hydrochloride, nicarbizin and arsanilic acid, meat and bone scraps, fish meal, vitamin B-12 supplement, poultry by-product meal (you eat your sister), dehulled soybean oil meal, corn gluten meal, wheat standard middlings, dried grain and whey fermentation solids, dried extracted streptomyces fermentation residue, riboflavin, calcium pantothenate, choline chloride, niacin, animal fat (preserved with butylated hydroxyanisole, citric acid, propylene glycol), menadione (source of vitamin K), Vitamin E supplement, butylated hydroxytoluene (a preservative), cod liver oil with added vitamin A and D concentrates, calcium carbonate, dicalcium phosphate, salt, potassium iodide.

Of course, if you want to raise caponettes (desexed males), inserting a capsule of the synthetic female hormone diethylstilbesterol into the necks of the chickens does the job. Heaven only knows what it does to the persons who eat these capons.

All I can say is: Not at our place. NEVER at our place.

RURAL VILLAGE SABBATH 🌿 JULY 1984

A grand old oak still stands in the midst of the fields behind our house. For two hundred winters or so it has stood there, stark, bare, and beautiful. Years ago a dear, now-departed Danish friend would take his chair outside on a winter's day and, all huddled up, look long and lovingly at its rugged form. To him the spirit of the tree was revealed more fully when the leaves were not there. His face shone after a session alone with his tree.

That monarch oak, now so fully clothed, watches over a host of animals and birds. Creatures fly or run for its shelter. Once they have gained its shade they simply disappear among its thousands of leaves. What a marvelous tower of strength and safety it is. The Indians camped nearby, if one may judge by the arrowheads found after plowing. How great it would be to be able to go back in time to see some of the sights this tree has seen. Think of the Indian families, with the children playing about, as they fished in Penns Creek. Imagine the roaming buffalo herds, the early pioneer settlers, the clearing of the fields. Can we conceive of the labor and time spent in laying with primitive tools the many rods of deep stone drains that criss-cross the fields? Just wonder at the variety and volume of the crops that came from the surrounding acres. What enormous piles of grain and grass and hay two hundred years would have witnessed coming out of the soil. An endless source of wholesome sustenance lies there, waiting to be taken and used. A never-ending marvel are the soil and all its workings.

Recently we journeyed sixty years back to the place of my childhood. One Sunday morning, following the service in the little old church in the wildwood, where my father had been the pastor for a number of years, we went to see my old home. We were seeking a maple tree that grew at the edge of an old canal just behind our house. Here I had learned to swim, to ice-skate, to paddle a canoe. Memories rose out of the very earth. Things appeared so much shorter and smaller and closer together than I had remembered them. The maple tree was now a declining giant. One of its limbs, which had hung out over the water, had recently broken off to lie on the bank near the base of the tree. Thereby hangs a tale.

As young lads we had tied a rope to this very same now-severed limb. Oh what a joy it was on a hot summer's day to grasp the knot we had tied in the rope, pull it far back on the bank, then swing out gloriously over the water. The rush of the air, the thrill of the drop as one let go at the far end of the swing, the cooling dash into the water, the swim back to go at it all over again—these were a part of a childhood of wondrous memory.

One very hot Sunday morning, in our best clothes as usual, we all went to our church, less than a quarter of a mile from our house. As father discoursed endlessly, our young minds and hearts wandered. Mine ended up, as the final "Amen" drew near, with visions of that tree, that swing. Sunday swimming was forbidden. One read quietly, or rested, or talked with one's elders and friends, or went for a walk, but one did not swim or play baseball or other active games. There were no television or radio or automobiles, and bicycles were out of our reach. We did not exactly revel in frivolity. But one could still think of how good that water would feel tomorrow.

The service finally over, in my precious, hard-earned new suit, I broke into a run as I neared the house, far ahead of the others. Surely just one swing out over that lovely water before they returned was neither sinful nor harmful. No one would ever be the wiser, anyway. From the foot of the tree I looked up longingly. To my great surprise that rope simply reached out and dropped itself into my hands. Nothing else was possible than to back up. And then the devil made me do the rest. Oh what delicious glory, enhanced by just the hint of delectable guilt. All too soon it would be over, and sedateness would again be the order of the day. All the pent-up desire was half consummated as the swing reached its apogee.

Suddenly an unaccustomed laxness set in: things were not holding together properly. An age passed as my ten short years flashed before my eyes. For one long moment I remained suspended in mid-air, motionless. Could this actually be happening to me? Does retribution always come so soon? Was I only imagining that the canal was rising very rapidly toward me? Was the rope so limp in my suddenly perspiring hands? Impossible! I simply dared not get wet, that was all there was to it. Perhaps I had gone to sleep in church and was having one of those dreams of falling to which as a child I was prone.

My best clothes clung to my limbs as I thrashed toward the shore. The frayed rope remnant on the limb twisted slowly in the wind. With mighty squishing of muddy shoes leaving a sad sodden path up though the yard, I attempted to spurt to the back door. It seemed to take forever, what with feet and legs encased with lead. I managed finally, after another lifetime, to reach the house and climb the steps to the third floor before anyone else returned. They had dallied in glowing self-fulfillment, drinking in the beauty of friendship and of spiritual uplift. Little did they realize that their young relative at the very moment of their happy return lay writhing on his pallet in anguish and torment, both soul and body naked before the world, his Sunday best a sodden heap under the bed on the attic floor.

Finally I was missed, then sought, eventually found. Why was I in bed on such a lovely day? Was I sick? "Let me see your tongue and feel your forehead. Oh dear, why is everything so wet?" In body-racking sobs the catharsis took place. "Mama," I bawled, "the rope broke!" Shock and dismay flew over my mother's features when she finally accepted the incredible truth. Immediately, as always, the inviolable sentence was pronounced. Stay in bed the rest of the day, without dinner. No swimming for two whole weeks. But the grimness softened, the face suddenly turned away to reveal from the back a broadening, a lifting of the cheeks. In time, when the awe had subsided, titters arose from the lower regions as siblings crept up the creaking stairway for just one glance at the littlest miscreant angel brother. How that hot summer afternoon dragged endlessly as the voices of neighborhood children floated up into and around the room. One remembers some lessons forever.

Is it any wonder that to see standing there the very same tree that knew us well as children, to relive that special experience of more than half a century ago, caused sentimental welling-up within? It was sad to see my tree in the midst of its final decline, with large dead limbs here and there. That day I also learned that a former neighbor whose whole life was spent in the next house had very recently passed away. I felt particularly close then to life's continuity. As I stood with bowed spirit for one last moment before reentering the present, suddenly from far, far away, from deep, deep down, seemed to come a clear, clean sound, as if from out of the childhood of the race. It was the tolling of a bell.

Fall

*A*t Walnut Acres, fields of Queen Anne's lace, the busy wild animals, the harvested crops, and the somber mood of fall engendered by hunters bring thoughts not only of death but also of life.

DEATH AND LIFE ❦ DECEMBER 1965

I write today in a predominantly brown mood. The browns are every-where. True, they are intermingled with greens and yellows and blacks, but the feeling is brownish. One might almost think of brown as the color of death, for spent leaves and grasses and corn-fields give accent to the quiescent brown soil seen through and be-neath them.

And death is all about us today. Determined men, eyes strangely alight with intent, stalk the homelands of the wild creatures, the long power of destruction hanging ready in hand. For this the hunters have waited and planned. This is their day.

Some few timorous beasties will cower to safety to see another season. At worst they are still less subject to human appetites and fancies than are their more domesticated animal brethren. At least they are not molded as to size, shape, structure, and function into the current pattern. Is short freedom to be preferred to long, regu-lated life? Bunnies, give answer.

Yet the brown rabbit's urge to elude the tearing shot is the hope of spring. The fallen hickory leaf can look up understandingly to the bud that has pushed it forever off. The dead-looking grass hides the life at the growing edge, where soil and air meet. Let us have several consecutive unseasonable warm days and all over the fields the green will again be turned on.

There may be death at the fringes, but not at the heart. For the heart of life never dies. Had we for but one brief moment the vision to perceive, the mind to comprehend, the strength of spirit to bear a glimpse of the true nature and reality of the heart of life, nothing for us would ever be the same.

But we are such small beings. It would be too much for us. Such feeble powers of insight as we now possess lie mostly wasted and unused about our feet, smothered by the distractions and deadening influences of our age. Give us the warm domestic shelter of our cities and our comforts, our amusements and our toys. Not for us to be shaken deeply. Not for us to cower before the Great Hunter's piercing darts of submissive perception, utter concern, and rejoicing wonder.

How every square foot of earth must teem with life, perhaps to depths greater than we realize. One cannot even faintly imagine the interlocking intricacies of the life forms, great and small, throbbing ceaselessly there. Are they driven by a purpose, or drawn toward a goal? How utterly remote one can be from the very source of life lying but a short shoe sole away.

The culture of earthworms in their twisted cylinders of clay, the development of the seventeen-year locusts awaiting their long resurrection, the activities of the fine creatures, microscopic and submicroscopic—all seem bound up in a living web of purpose, of continuity, of determined growth that never dies. Perhaps the slightest wriggle of the meanest creature is a part of the whole pattern of our lives.

Yet today people treat and spray and poison and incite artificiality with furious activity. They plunder and rape the soil to its unbalancing. Isn't it just in order, really, to have more comforts, more amusements, and more toys? And to have more of these than anyone else has? Perhaps we think in the process to domesticate, to bring under our control, the very heart of life itself.

For life has a terrifying independence, a bewildering wildness that takes it ever determinedly on its own way. Aspects of it may seem to bend themselves to our will, but by and large life sets the direction, and we must follow.

We are learning. Daily grow the numbers of those who in their hearts glimpse a part of the secret: that one must work with life, putting back more than one takes out. One can think of this attitude

as being in concert with the soil and all Nature, in the true understanding of one's body, and in person-to-person association. People
knowing this secret become the true conservationists. Exploitation
falls from them. For them, truly, the whole Earth is under the rule
of the glory of which Isaiah was told.

Normally in the country a deep quiet pervades the night: no
sounds, except for the darkness breathing outside the window. Oh,
with a westerly wind the hum of a late truck sometimes comes floating across the fields from the mile-distant highway. Or one may
hear a lone cow, separated from her baby, exclaim with awesome
vehemence from a neighbor's barn. But mostly the night is for dark
and quiet and sleep. Overhead, from horizon to far horizon, the ir-
regularly edged, inverted bowl of stardom seems to hold in the
quiet. A violent intrusion it is when a mechanical beast, pistons
flailing, dares to steal across our heavenly concave, winking in
among the stars.

But last night was disturbed. Sound sleep was elusive. A lovely
new moon moving fast to full hung on its invisible sky string.
Bright, bright stars, each bursting to outdo its neighbor in radiance,
clustered around the bit of moon. But not too close, for moons put
out stars, you know.

From the old oak down the lane a moping horned owl did most
sadly "to the moon complain." As the deep resonant "hoo hoohoo
hoohoo" quavered through the frosty air, its note of despondence
did nothing to soothe our already troubled spirits. It quavered its
way too into the dull consciousness of the old hound in her box on
the back porch. Sleep-befuddled as she was, she could still send the
hushed air molecules scattering in all directions as she split the si-
lences with her age-hoarsened protests. Soon all the dogs within a
one-mile radius took up the cry. Canine bedlam reigned as the word
passed over the countryside. In the distance, long before dawn, a
cock crew in the general excitement. Peace was shattered. Owl, all
a-chuckle, left.

Sleep was gone. How about the men and boys of all races, creeds,
and colors hunting each other down in Vietnam? And how about
our relatives in Rhodesia—where will it all end for them? Around

and around and around it goes in one's head. The bitterness of it all. So much of human action is an attempt to cure after the disease has struck. Prevention is, like peace, so unglamorous, so hard to keep at. It is so easy to forget that the disease is often but the final symptom of earlier, prolonged imbalance. Perhaps again the world did not try soon enough or hard enough.

Despite the effrontery of signs to the contrary I believe that humanity is learning here, too. Generally speaking, perhaps, we are no longer proud of going to war, to kill and to be killed. We may think of it as a job to be done, but in our hearts we begin to suspect that we may have failed somewhere. We're ashamed at not having acted sufficiently upon the understanding we already had. Some day war will be beneath us because we will have risen above it.

Now the year will soon turn. The days grow shorter and shorter, their clock running down. Something is about to enter the picture to reverse the proceedings. We know a bit about the "how" of this "something," but we don't know the "why." Long ago by our clocks, far away by our rules, shepherds lay by night under their Judean star dome. They longed to see the hand behind the stars, to know the "why" and the "something" that held them together. And glory be, they did see!

We wise moderns may scorn this old-fashioned story from simpler days. But perhaps we shall feel the mystery of the "something" about us with especial closeness at this season. For us, too, the angels will sing. The "whys" shall be ever more important to us than the "hows." Perhaps through it all we hope to become persons of goodwill, to whom peace will come at last on this Earth.

FALL PLOWING ❀ JANUARY 1985

Fall plowing is the most marvelous eraser. As always, summer-crop fields lost something of their sparkle once the harvest was over. Tomato fields, scruffy with decaying red and green fruit, plus dead, vining stems, had become a sorry sight—a sad end to so valiant an effort. Weeds had begun to come in everywhere, once the soil was no longer shaded by the plants. Let the healing plow pass but once over the field and the humpy field's slate is suddenly clean. Gone is life's debris, buried are life's remains, turned over is a new leaf. Some fields are "left open" all winter. Roughly plowed, they catch and hold the snow, helping to keep it from blowing about, allowing the occasional warm-spell melt to sink more readily into the loose-soil reservoir. Further, the head start the farmer gains in the spring from fall plowing, particularly in a rainy period, can make a great difference. A roughly plowed field presents infinitely more surface area of soil for sun and wind to suck from. Thus is kept going the endless, life-assuring tussle between Earth's seeking greedily to draw everything to itself and the sun's resurrecting power struggle to liberate ever again Earth's water supply for a large, grander, life-giving purpose.

Still further, if one can plow in the fall, crops may be planted earlier in the spring. Each acre holds millions of miles of roots and rootlets and root hairs to be magically changed from being feeders themselves to being food for new roots. When wet spring comes, most of the requisite rotting down of the sod will have taken place through the marvelous alchemy of the soil stomach. Fall plowing also permits us to make tight, close soils like heavy clay considerably looser, more friable, more easily worked, more ready to give up their substance to the following crop, if they have been exposed over winter to the heaving, kneading, stretching, and shrinking action of freezing and thawing.

And so, come a spring dry or wet, fall plowing can be helpful. Providing that it is not too wet in fall to permit heavy machinery in the fields, of course. Or that the field to be plowed cannot wash readily. It is foolish to risk losing precious topsoil by having steep, hilly fields "open" in winter. Of course, such fields tend to dry out earlier in spring than low-lying ones, and they may generally be plowed earlier. But with fall plowing one cannot plant a cover crop both to hold the soil over winter and to provide soil food in spring when plowed under. How many decisions must enter into a year's work.

Some crops must have residues from former crops already well digested, or they will not do well. Plant field corn or other heavy feeders in freshly plowed and heavily manured sod ground, and it will tear its heart out to grow and produce; but plant a more delicate, light-feeding crop in the same manner and it may fail. Such crops must have the heavy burden of organic matter well digested first.

Nearness to water must be considered, too. We must plant most vegetable crops in fields near the clear stream of Penns Creek, in case of a prolonged dry spell. Otherwise, without sufficient water, they may stand still or wither away. The grain crops, planted in early spring or fall, usually have enough moisture to go on, even if rain is sparse. And so they, unpampered, get the higher fields, far from irrigation.

Finally, here in the northern clime we find our farming selves squeezed inexorably between the frost's icy jaws—its early and late limits. One never knows when these will clamp onto one's fields and crops. The midsummer may be gay and carefree, but it is the growing-season extremes that give us pause and sometimes worry. Whole crops can be lost if the corn is still soft when the frost hits, or if the soybeans have not changed from a spongy oblong shape to a firm round one in the shrunken pods.

Nothing is new in all this earth knowledge. These concerns have been those of the people of the soil since life began. But it may be helpful to have all the wisdom reviewed occasionally, and particularly now, during the shorter days, while we sustain ourselves on the harvests of the latest growing season. Only good can come of calling to mind once more some of the down-to-earth details from the incredibly involved and intertwining system by which breath is sustained. How often we take it all for granted, as something that life somehow owes to us for our submitting to being born, to grace the universe for a time with our presence.

At the heart of it all stand the farmers. Thank God for them. Their first care lies in preserving the soil into an endless future. Too often they are grossly underrated as bumpkins, the man-with-the-hoe type, brothers to the ox, too stupid to move to the city where culture and affluence really reside, misfits of the postindustrial age. On the contrary, farmers must have a wit, a wisdom, and a flexibility that can put to shame those of many urban compatriots. Who else can daily read Nature's book so well and dance so nimbly to its every tune, its slightest frivolity?

THAT FEEL IN THE AIR ❦ OCTOBER 1959

Just now, outside the cabin in the woods, a crow gives one of its end-of-summer calls. Close eyes and nose, lest these betray the season, and listen again. Ah, but there's that feel in the air. Is that what makes the crow sound as it does?

A perfect, quivering spider's work of art is anchored to those bushes just beyond the open window. Now the first shaft of early morning sunlight flashes across the upper half of the web. The silken cables turn to shimmering fire. An enormous dew-pearl, earth beckoned, slides from pokeberry leaf to spoke, from spoke to center, thence to its destination. The power that calls the dew to its home also shapes the spider's art into a drooping exquisiteness that defies words. The slant of the sun's rays, the time of their appearing at this spot—these speak in end-of-summer phrases. Or do we add these, somehow unknowingly, to that feel in the air?

A grasshopper, life's summer behind him, comes from nowhere to clutch desperately the stark, swaying timothy stem, there by the goldenrod. The seed head is gone. Its future lies about it in myriad promises of life continual. Even now each of the progeny must feel a strange stirring within, a response to the eternal call of the good earth to give allegiance and to gain life. Soon tiny rootlets will begin to explore the covenant.

The crickets know it. They plan their winter shelter ahead. They invade the house in droves. An army of them welcomes every door-opening with the most deliriously ecstatic leaps you ever saw, twisting and turning as if utterly unconcerned, but always keeping an eye on the ultimate destination.

Squirrels too feel something in their bones. Early in the morning every nearly ripe hickory nut is inspected, with time out for animated discussion. Let the outer shell indicate just the right degree of looseness, and off it comes. One more for the larder. What chance have little boys and girls against this drive?

The soybeans have run through all shades of green. Now, just now, they begin to yield to the golds, the yellows, and the browns. They have had a gay, carefree summer, befriending rabbits and woodchucks, and perhaps a deer or two. Next come the leafless stems, bean pods hanging with true Oriental grace. The full corn in the ear. Then the harvest.

The wheat ground is worked. Very soon the new turn of the wheel will begin. Again the seed will be planted. Take a handful of that soil, or stretch out flat with your face over it, and draw in life's odors. Can you explain what the smell of living earth does to you? Will the agriculturists give us a formula for it? Ah, the completeness, the synthesis, the balance of it all. These will, thank God, forever and ever defy analysis and elude description.

Framed by our west window, the village peeps out of its dreamy past. Old-fashioned in a way it is, with no particularly notable features, yet peopled and lovable. Atop the village stands the mountain, and atop the mountain "the infinite, tender sky." Fields, trees, smoke curls, rooftops, church spires—all blend in the diffused early morning haze before the backdrop of the high, wooded hills. Truly, in this light is the valley exalted, the crooked made straight, the rough places plain. All Nature has burst forth into singing. We need only the ear to hear, the heart to understand, the spirit to praise.

Summer has come and almost gone. It has been wonderful. Crops were good. Soil tests show the life of the soil, the organic matter content, the available mineral portion as constantly on the increase. Grain tests show constant improvement in quality. And now for a slow, relaxed, rewarding, joyous fall.

HARVESTING CORN ❊ SEPTEMBER 1982

Soon the combines will begin to roar through the fields of ripened corn, chewing up four rows at a pass, moving at a pace that would keep you stepping to stay with them. Belts and pulleys, gears and axles whirl and spin. The fifty-thousand-dollar machine can pick and husk and remove corn grains from perhaps two hundred ears a minute in good going. That's a lot of money to have invested, especially if it isn't one's own. One uses this monster for several weeks out of the year to harvest all the grain. Then it becomes the winter home for chipmunks and other varmints, which seem to thrive on rubber belts, pieces of canvas, and other rodent delicacies.

Recently one of our big tractors, having run more than five thousand hours, developed low blood sugar, or something similar, and had to go to the repair shop. The shop owner just happened to have a brand-new tractor that he wanted to have broken in, and so he kindly let us use it—which we did, for sixty-five hours. Before trying this new machine we had thought our wide-open, fresh-air beast was still new. But this temporary replacement had it beat hollow. You sat in a luxurious, enclosed, soundproofed, air-conditioned cab, with grossly overstuffed seats, an AM and FM stereo, and a tape player, plus enough knobs and buttons to entertain a two-year-old for a whole afternoon. At thirty-five thousand dollars it was a steal. Yet at a sale the other day a beautiful ten-room house in perfect condition in a nearby town could not bring more than twenty-three thousand dollars. Needless to say, we're back in the saddle of our repaired tractor once again, happy and rejoicing in its simplicity.

Here comes that remembering. As I look out of my window across a wide field to the "mountain" beyond, my eyes glaze over, and in a trice I see Molly and Prince, the calm and steady Belgian team of horses I used before I got the tractor. Those two hayburners were real personalities. They cost about five hundred dollars, and they lived off the fat of the land. In those days the farm was nearly self-sufficient in energy, except for the putt-putting old stationary gasoline engine used to saw wood, shell corn (remove it from the cob), and grind feed. Our hands were the milking machines. A hand-turned clipping device sheared the sheep. Nighttime barn work was done by romantic kerosene lamplight. Trees were reduced to manageable logs by hand sawing. Yes, horses and men did the brute work.

We could not farm the soil as hard in those days. Life was simpler. We didn't want or ask as much. There weren't as many things to want, as many ways to make one want them. We trusted life and our own abilities far more. Each person "made do" for himself and herself, leaning, by and large, upon their own devices. Those who didn't want much didn't have much. Nowadays we all want much just because it's there, sometimes free for the asking or taking. Sometimes people are even almost begged to take this or that handout from the common purse, please. How strange things have become.

When corn is picked, husked, and shelled by the combine now, the grains contain about 30 percent moisture, and if stored immediately in a bin would mold in no time. The grains must therefore be run through a dryer, fired by propane gas, until the moisture is down to 12 percent or so. Pass by a dryer in fall and watch the steam coming out. Of course the higher the heat, the sooner the job is done, and the sooner the life-giving germ is destroyed. To preserve the corn germ, we dry ours slowly at very low temperatures.

In our youth we had two ways of harvesting the corn. You could cut the corn, stalk by stalk, with a large knife until you had a big armful, which you then tied into a sheaf. A number of sheaves made a shock. The corn shocks stood about in the field until, stalks disconnected from the roots, the ears dried down some and were easier to husk. Sometimes in the fall these shocks would be gathered into several vast master shocks, to save moving around when the husking began.

Then one luscious, crisp, infinitely bracing and beautiful morning, a husking bee would begin. It was little fun unless the company was mixed. Maybe the frost was already on the ground. Jackets and sweaters were usually needed until ten or so of a morning, after which the clothes-peeling began.

For husking the corn, each person had a husking peg to hand, a hooked piece of metal fastened by a thin strap to the hand or to a finger. One jab with the bent-over hook bit through a portion of the husk and silk at the top of the ear. By then grasping this loosened clump of husk between peg and thumb and forefinger, one could pull down from the ear a portion of the husk, exposing the lovely rows of gilded kernels. What a beautiful grain. After that the rest of the husk could be torn off, although often an overall toughness required one last twist to separate the ear from the husk.

It all went remarkably fast. One developed an incredible economy of motion, and the husking soon became automatic. Joyous banterings and badgerings soon filled the warming air. One tossed the finished ears onto a central pile. Others would then come through the field with horses and wagons to pick up the piles of corn, later to shovel them from the wagon, throwing each shovelful up into a high door at the top of the crib. The stalks were gathered later, brought into the barn, and chopped for fodder and bedding.

Sometimes a handsome, deep-red ear would appear in one's hand. The girls often hid theirs, but the boys whooped and hollered and claimed their rewards, because a red ear was a passport to the kissing of each girl in the party, and it was generally hailed joyously. Some boys were known for their dexterity in squirreling away red ears from the pile without anyone's notice, and they fared somewhat better than the others. Husking times also brought a warmth and richness, a chance to philosophize, to recite, to sing, to promote fellowship, which one misses in these labor-saving days.

Let the thought not be harbored that all this harvest-time activity was only fun and games. It took perhaps thirty to forty people to do the work that the one giant harvester today picks and shells. Cutting stalks of corn all day long with a kind of machete called for very tough skin and hands, and hardened muscles. Husking soon brought out the best in a person. Hands could be cut by the husks, skin worn off by the hard kernels on the cob. Picking ears from the ground to throw into the wagons was a backbreaking job. Did you ever try shoveling hard ears of corn from a full wagon? After you'd reached the flat bottom, things went better. But to toss heavy loads high in the air and hit a small crib opening was no easy sport. Any that missed the doorway often bounced back on one's head or body. But at night, head had hardly touched pillow before blissful oblivion and rebuilding set in. After a week or so one became toughened and took everything in stride. The fullness and satisfaction that came with it all is difficult to describe but easy to remember.

The other harvesting method was first to tie muzzles over the mouths of the horses, then to hitch them to a wagon that they drew through the field of uncut, standing corn. The first couple of rows were specially husked. Then the wagon was pulled between rows of husked corn. A large door was placed lengthwise inside the wagon, along one side of it, to intercept the thrown ears of corn. Here we husked ears from the stalks as we came to them, hurling each ear through the air at the door, generally hitting it. This method saved a number of steps, but the stalks were often not saved for fodder or for bedding. They gradually fell down, to return to the soil.

How often babies slept in baskets at the fields' edge, while mother and father husked and older siblings scrambled about helping as age and condition permitted. It was a beautiful family venture, readily understood and accepted by all as contributing to the necessary work of making a living.

The modern harvester surely saves time and labor. But it has changed many things. One sometimes has long thoughts about it all.

HUNTING ❦ DECEMBER 1968

Just now a lazy-looking hawk slid along the mountainside at close range, as if covering miles in seconds. One cannot help marveling at the contrast between effortless coasting in billowy air and the piercing power of that earthward gaze. Wild, hidden, tight-wound energy comes to a focus through that eye.

Let a field mouse, a hundred feet below, dare so much as to step outside its grass tunnel and ecological forces will act with incredible speed. Small wonder that the little creatures thrive so in the unshorn grass covering large parts of our farm. We're told that more small game is to be found here than on any other farm in the area. Pheasants and rabbits and squirrels abound. This plenitude places us in just one more of the world's dilemmas. We shelter the wee beasties so that more of them may be hunted down.

Volleys of shot have clapped out all morning along the same mountainside. There other ecological forces act like lightning as the soft-eyed doe takes her last long leap into the searing bullet. On many a wooded slope, leaf-brown, limb-shadowed, the mother deer will be harvested on this bright day.

Two black bears will not see the light gray winter clouds scurry across the hills today. One of them ran fatally into a moving vehicle just a mile or so south of us. The other, after raiding neighborhood garbage pails, was dispatched not far away during bear-hunting season.

A red fox sprinted by the window a moment ago, losing himself in a hillside grove of young evergreens and thick brush. Years ago we planted these trees, supposedly to help put the girls through college. How fast the years, how slow the trees. Four or five dry years have not held back the girls. The first is out of school and two years married. The second plans to graduate next June, and the third and last is well started in her first college year.

A later generation will lose itself in the splendor amid the tall, robust spruce and hemlock on that hill, dreaming its own fresh dreams, thinking little of the hopes and joys that went into the planting, accepting the trees as having been there always, little knowing that at one time grandfather (or great-grandfather) held each of these trees in his hands, root and all, just as he held their own mother (or grandmother). If only wooded slopes could tell what they know of trees, girls, foxes, and hopes.

Any night now that we drive down the old lane, we must be careful not to undo the opossum family living in the steep bank below the road. Almost invariably one member of that family will be foraging nearby. What a lot of slow, stupid scurrying follows as the headlights work up that long nose to those dimwitted little eyes. We once found a possum in the barn nuzzling the rich alfalfa hay, much to everyone's mutual surprise. We were reminded of the old cow we lost, and searched for all over the neighborhood for the better part of a day, only to find her too, at last, nuzzling the hay on the barn floor.

One of the neighborhood boys went on his first raccoon hunt last week. Oh, boys! Toddlers only yesterday, they have suddenly stretched their thin selves into hunters' boots. With nervous self-confident nonchalance they take first gun in hand and disappear, red-coated, into the trees. At dusk home they come, proud initiates, one important and meaningful adolescent hurdle passed.

One must wonder if all this hunting fever is but a prelude to the hunt by man for man. Directly beneath our feet, as the worm burrows, men stalk other men today. For a few hours at Christmas, in deference perhaps to some hazy notions about the Prince of Peace, the guns will cool. Love and brotherhood will surmount all defenses. Then back into His box goes the Prince, enemies will again address each other with shot and shell, while at home ear and eye will daily inform our callused, numbed souls about the number of world families torn that day, and bereft, by the world's bullets. How long, oh Lord, how long?

How hard it is to live in a small world when a larger one lies in view all about us. How difficult to pretend that the daily foot-dragging journey along the low road fulfills, when the high road ever calls from above. Must the ideal be forever so far beyond the reach of the real? Can the visionaries ever build a bridge between their two worlds?

Yet despite the cruelty of it all, life does go on. Leftover bunnies still hop quietly across the lawn, poking fun at the old dogs snoring in their boxes. Pheasants, subdued now, the flush of life passed for this year, shelter in the fencerows. The year has quieted down. Its end is approaching. It too was born but to die.

But life itself skips among the years. Tall, dried bromegrass stalks at the field's edge oscillate in the breeze, their seeds long since sown. Just beyond them lie a billion new wheat stalks, greener than green against other browns. Out of life, then, death. And in the midst of death, life.

JOHN F. KENNEDY ❦ NOVEMBER 1963

Rain falls incessantly today. Silver streaks slip silently past the eye, past the window, their final step in the long journey from heaven to earth. A volley of shotlike rain rattles on the old tin roof as a moisture-laden breath weaves through the somber branches above.

All is bareness now, save for pregnant pearls clinging precariously to branch and twig, each awaiting the final plunge, the birth of another drop.

A mist pall hangs brooding over field and stream. A fencerow tree on a far horizon stands espaliered against the gray gloom. The mountain melts into the mist. At its foot the village weeps.

A CAISSON, WHITE-HORSE DRAWN, RUMBLES STOLIDLY ALONG PENNSYLVANIA AVENUE

Yet the earth is made to be wept to, and the sky is made for weeping. For rain and mist do not live to themselves alone.

Far away in the west, warming the topside of the cloud blanket, the unseen sun smiles on at work well done. Once more for us it has lifted the life-giving vapors from a thousand mighty reservoirs, so that again the thirsting land over a thousand fields might be refreshed, revitalized.

And so heaven's tears become earth's boon.

And the cycle goes on forever. From land and sea, sun calls his handmaid skyward. Then follows a spreading out, a dropping down —floating flake or air-molded globule. What consummate artistry! And once more earth's wrinkled countenance accepts the cleansed gift. Oh blessed rain that fosters life. Oh hallowed gloom that cradles hope.

And once more, too, perhaps, for a time the vital fluid enters into the unfathomable mysteries that spell life and growth and death to all things. But for a time only—for an hour or for a thousand years, it matters not. The time value is not earthbound.

The fluid may appear in lowly form, as part of plant cell or soil bacterium, of paramecium or protozoan. Or it may fulfill its highest possibilities and help to make a human being. To think that this priceless fluid can lend itself to the profoundest dreams that ever stirred in a noble human mind!

But whether here or there, after short time or long, in base form or noble, to it at length the recurring summons comes. Unheard by ears attuned only to coarser sounds, the gods speak.

The long call resounds in the hearts of a thousand thousand drops. Their parts done for a time in the eternity that is life, they leave their tasks, drop their burdens, free themselves from earthly entanglements, and return to sea and sky. And life withers.

Oh, it is hard in the present state of our inquietude to write this account. Each word is a struggle, each idea hides behind a shadow. The too-easy sentences fail to come now.

A spiritual hiatus settles in. We move as in a dream. We go our daily round but find our hearts elsewhere. Or nowhere. We are a shaken people. We now know that our strength is insufficient, that our destitution shows through our trinkets. We are pretty sorry captains of our souls.

If rain and mist truly cradle earth's hopes, can our spiritual gloom, analogously, nurture the soul's aspirations? Are life's dark days preludes to a growth in understanding, to a maturity to which without them we would never attain? Can we be asked, even freshly wounded, in the overwhelming flood, to see through to a light peculiarly our own?

A BEAUTEOUS WOMAN, WITH BEAUTY ENHANCED A THOUSAND-FOLD THROUGH COURAGE AND FAITH, STANDS VEILED IN A NATION'S HEART, THE FUTURE HANDCLASPED AT EITHER SIDE

This year we need Christmas as never before. We need the assurance that life is in its completeness so much larger than we are, that there is so much more to it than the sum of all its drops and grains and cubic centimeters.

We want to be told again that while a spirit draws together, guides, and then disperses, along with it are both purpose and perpetuity. We want to feel God's touch upon us. We are children, and now again we know it. Too long have we stood thinking to hide the heart's whimper.

More than nineteen hundred years ago in a tiny land stood gloom and dejection, sorrow and sighing. Life was hard, seemingly purposeless. And then one night darkness fell and stars shone out and shepherds watched. And of a sudden the veil shattered and heaven invaded Earth and angels sang. And a stable was hallowed. And here we are, strange creatures into whose hearts creeps a warmth and a glow and a rising up. All because of a star in a tiny land.

This Christmas the glow must be a dross-consuming fire. Nothing less will do. And next year must find us wiser and better and more worthy of this wondrous life.

LAMBS IN TWO WORLDS ❋ JANUARY 1961

The somber, still sheep, noses to the ground, continue to wander about the pasture, nibbling hopefully the frosted grass. Not enough food value can now be found in a day's foraging. With many an eager "baa" they call for attention along toward evening when the school bus brings the girls home. The hays and the grains are quickly gobbled up. The trough is licked clean. What intensity animals put into their licking. In old horse troughs one finds walnut planks worn half through, one wood molecule after another yielding to the great rasping tongues.

Pooh and Eeyore (Eeyore's ears flopped in her youth) were twin female lambs born last spring. Their mother wouldn't let Pooh nurse, and so we raised her on a bottle. Eeyore was caught by a dog and half skinned alive, but grew new skin and wool, of all things. In being cared for by us she became a part of our human family. These lambs are members of the small flock now and are not half as tame as they were. But when they look at us knowingly, out along their black noses, we can easily read into their glances a kind of half-human recollection of days when they depended fully upon us.

Their daddy was a small fellow who never grew very large. Stunted by both nature and circumstances, these two lambs are the Lilliputians of the flock, but what they lack in size they make up in character, for they have lived in two worlds. At night, as they lie folded with the others, in dreams they may slip from one world to the other. They are babes again.

MAKING APPLE BUTTER ❋ JULY 1982

In fall 1946, our six old apple trees produced their first crop for us. We were amazed to find the fruit practically insect free. We felt sure that those trees had never been sprayed. Question: what to do with ten to fifteen bushels of precious apples, our first, our very own?

Apple butter had always been a part of wintertime fare in my Pennsylvania Dutch background—a way of saving the fruit for late, delectable use. At a farm sale we had purchased a kettle and tripod. What better than to make our first lot of apple butter on this, our brand-new farm?

We were already cutting our winter wood supply (by hand, no chain saws) because we used only woodstoves then. And so we had everything we needed but a recipe and a fancy name. We tried making small lots from varying recipes until we found one that seemed the very best for our mixture of sweet and tart apples. We used no added sweetening, of course, but found the result so delightful that we have used the same recipe ever since. And the name? Well, because we concentrated the apple flavor in all its deliciousness by boiling off most of the water, we finally settled on Apple Essence. For many years this was the product's official name. In that first year we made a hundred quarts and sold them at a dollar each. How that helped the early coffers—and it also helped us consider seriously the idea of mail-order foods.

As this is written, our thirty-sixth "apple-essence" season is in progress, but with a difference. We now boil it down with steam, in indoor kettles. The open sky, the scent of leaves, the smoke are gone. We really cannot detect any difference in the wondrous flavor, but our hearts know the difference. Making apple butter in the new way does not do for us what the old-fashioned way used to.

Here is how we used to do it. Some time before the making, firewood was gathered, split, and stacked for use. Some of the apples had to be made into juice, and so they were taken to the old cider press, where one waited hours for one's turn to get to the business of pressing. Valves and pistons clicked and whistled and squealed on the old one-cylinder gasoline engine that powered the whole complicated system. The more power required, the more the engine chugged. In periods of slack demand, the huge flywheel kept the engine loafing along with only an occasional "putt-putt" to maintain a minimal speed. When power was called for, each explosion of the compressed fuel could be counted, as it burst upon the scene, denoting its tremendous power, background music to a world of pleasurable fulfillment.

At the apple press, children were everywhere. Young maidens and swains cast long sideward glances at one another, shy smiles lighting up blossoming faces. Horses and wagons were tied to all the trees. Bees and yellowjackets had a field day with the pile of apple pomace that remained after the pressing. One suspected that stray insects found themselves squeezed into cider from time to time. Happy banter, smiling faces, philosophical oldsters, a country-fair atmosphere, all made one almost sad to leave when the apples were finally pressed, the heavy barrels full of the most delectable apple juice.

When one's turn came, one dumped the bags or baskets of cider apples into a trough for washing. They were then carried up by an elevator to the top of the press. There they fell into a grinder, which chopped them into bits with a roar, accompanied by much putt-putting and considerable flying about of juice and pieces. The mass fell onto a sticky, soaked square of canvas, held up by a sturdy wooden piece, all surrounded by a wooden form. When several inches of shredded apples had built up on the canvas "pocket" within the movable form, the mass was moved into the press, where tremendous pressure was applied. Again the putt-putting.

The juice leapt from the canvas, running down in all directions to a hopper and then flowing into one's barrel. Three gallons of juice to a bushel of apples was average. When the mass was squeezed dry, the press opened, the pomace was dumped from the canvas, and things began all over again.

Next we prepared the applesauce, first cutting up the sauce apples, cooking and straining them. Then we combined some sauce and some cider in set proportions and cooked and cooked all day long, stirring constantly toward the end, until the proper texture and color were reached. Finally, when finished, the apple butter was placed in the jars and sealed. We often finished late at night, after the little ones had long since taken to their beds. How those jars gleamed on the cellar shelves.

I'll never forget the contributions of Betty's father as he held a baby on his hip, in one arm, cooing and singing, and used his other arm to work the stirrer. He spent his last years with us. Surely he stirred something into the souls of those young ones as he completed our joy in the day's work. Remembrances are sweet. Something in all this speaks of the way in which life was meant to be lived.

QUEEN ANNE'S LACE ❋ SEPTEMBER 1974

Have you ever examined closely a lovely, whitish or pinkish flower of Queen Anne's lace? Try it some time, looking first at the incredible top, which presents an open, smiling countenance to the sun. Its upper surface, sometimes convex, sometimes concave, is made up of perhaps fifty to sixty miniature copies of the whole. These subflowerets are shorter-stemmed and more compact as they approach the center by which they are held together. They stem from a common, green pincushion head.

Each stem arches gracefully to hold its own small group of flowerets in exactly the right position to contribute beautifully to the glorious effect. A lacy green upholding elkhorn collar grows from the base of the stem-cushion.

Out of somewhere close to the center of this stem-cushion often grows a special stem that bears a kind of purple-heart floweret. Sometimes this part grows alone on its stem; sometimes it shares its place with its common white relations. A long time ago a purple dye was made from these little blobs of color. When this stem is crushed between one's fingers, the color is released.

Almost a hundred years ago my father as a lad, along with his friends, picked many of the purple flower centers to be sold for a few cents a pound for use as a dye. How many of these delicate color-bearers, so shyly tucked away in the surrounding forest of whiteness, are required to make a pound? How many of today's children could bring themselves to pluck them at any cost?

The underside of the whole flowerhead is a picture of arching grace and symmetry. A shower of miniature green stem-cushions sprays from the central head, like sparks from an exploding aerial fireworks display. There they are again: miniature elkhorn collars, miniature stem-cushions, miniature arching stems, to make fifty or sixty miniatures of the whole head. One wonders if the replication goes on to infinity; one longs for magnifying eyes.

As children we were wont to pluck the complete flowerheads, leaving a long stem attached. We would place the stems in ink, some in blue, others in black or red. How interesting it was to watch the individual tiny white petals color up. Today I show my grandchildren these same wonders, finding myself transported marvelously backward through time.

As decorations for the form of the fair Queen Anne lay in the mind of her lacemaker, does the dream of the pattern, the design of the flower reside in the "mind" of the flower? Has it been carried in the seed from the dim mists of time—and will it remain there until the end of time? Will we ever touch and measure it? Do we need even to search?

No heavenly scent is exuded by this lowly flower, although over the years one comes to associate its odor in one's mind with early fall. Ants seem to love it, either because of or despite its strange perfume. But, oh the sweet secrets betrayed by its form and grace, its vitality and strength; oh the beauty of the commonplace. When connected to its parent Earth it is the essence of organization and tenacity. If not cared for properly, whole fields will be taken over by Queen Anne's lace, a sea of whiteness come August and September. Nature abhors a vacuum. Queen Anne's lace is ever ready to fill in.

But pluck the flower and leave it out of water: very soon it becomes limp, sprawling, colorless, and even odorless. Determination and drive, a short while before so strong, lies now forever unfulfilled. Once separated from its roots, its hopes and purposes fade with life.

One must think of the milling millions in today's world who have lost much of this essential contact. How wonderful to be living in a time when the pendulum is beginning its return swing. Once again is vindicated the faith of the true men and women of the soil, who may be blessed at least in part with ability to see life whole, and who live up to their best insights despite all.

PAST AND FUTURE ❦ FEBRUARY 1986

It all began at the farm of a friend in the Catskill Mountains of New York, in summer 1940. Betty and I had been married for only a few months. We had just returned from the teaching stint in India. I had never quite recovered from an attack of a virulent tropical disease, and we hoped a summer on the farm would help. It did.

Something about Bill Simpson's place lit fires that have never since died out. A seminary graduate, Bill had at one time been, even in the winter, a barefoot street-corner preacher in the metropolitan New York slums. Then, completely rejecting the ways of the city, with all its artificialities and glaring inequities, he found himself on a poor, isolated mountain farm, miles from anywhere, alone with the soil, the sky, and his thoughts. Gone for him the rich-eat-poor, the city-eat-country concomitants of urban life. Heaven had been gained at a single bound.

As a student-builder-farmer, Bill was ideally suited for gathering and applying ideas that interpreted the timeless for the present age. A tremendous hilltop library developed, tying him with golden threads to all the world. What thousands of young people would give today to find themselves in such a never-to-be-repeated situation as was ours.

We had freedom of garden and house. There I fasted for ten days, drinking only water. There we came to love the soil and the working of it. There too we came across a number of how-to pamphlets written by a former Columbia economist, Ralph Borsodi, with the help of his wife Myrtle Mae. From these we learned that in Suffern, New York, was a place called "The School of Living," founded and carried on by the Borsodi family.

We knew little of gardening or farming or food preservation at the beginning of that summer. By fall we had learned much. It didn't seem quite fair for us to go back to teaching mathematics and physics for one more year. Whenever I should have been working on a doctoral thesis, before my eyes swam visions of fertile fields and growing crops, of barns and animals and small, tender, living things. My heart belonged now, in a way both exciting and calming, to another world, at the doorway of which I stood awestruck. It was hard to finish out that year of teaching.

We had obtained some of the how-to pamphlets and studied them avidly. We knew how much we had saved by canning our very own tomatoes and beans and rhubarb and raspberries. We began also to learn the secrets of grinding wheat and baking bread. Because we lived only a short distance from Suffern, we found ourselves making an occasional trip there to talk with the Borsodis at their school.

Later, when other friends returned from India and stayed with us for a time, we all went together to visit the school. They were fascinated with the whole concept. They had carried on some of these practices on their own in their mission work. Upon receiving an invitation, they decided to stay on as codirectors of the School of Living. By the end of that school year we too knew what we were destined to do. And so in spring 1941 we also moved to Suffern, as co-codirectors of the school.

Our two years there were rich and formative beyond all asking. In the air were some of the first faint stirrings of a desire to apply decentralist principles seriously, of back-to-the-land movements in the form of cooperative communities, and of a larger questioning of people's proper relationships to each other, to society, and to the soil and all natural life about them. Even in those early days, the easy assumption that humans were the undisputed masters of all they surveyed was seen by persons working at the growing edges of life to lead to imbalances that could never ultimately be righted by any combination of humans' assumed cleverness alone.

Ralph Borsodi had drawn together, mainly by the force of his example and eloquence, a board of directors that included visionaries from among church groups, university faculties, political parties, and financial institutions. When these persons gathered, the thrill of a new hope could be felt coursing through their deliberations. Here were people pioneering their way into the future. Those of that early group still living today must view the present scene with knowing smiles and not a few chuckles. It is good, so long before, to have seen into the future and to have shaped one's life to one's dreams. To know that earlier one has not been all wrong can generate occasional warmth and glow.

The positions of Ralph and Myrtle Borsodi at the center of things were fully justified. They had built with their own hands a lovely, imaginative stone house, and the community of homesteaders had done likewise. They had gardens and small fields. The main school building had its satellite sheds and barns. They had their looms, their mills, their butter churns, their cheese presses.

At the heart of it all breathed the library. Incredible accumulations of ideas, for many of which the time seemed to have come, were squeezed within its four walls. Often the library's concepts seemed alive, moving through solid matter, coming to rest upon all who passed through the school. And there always, to challenge, to interpret, to direct, to expand, stood the Borsodis—original thinkers, indefatigable doers, born teachers.

There we both learned and taught. Composting, gardening, milking, butter and cheese making, grain grinding, bread baking, beekeeping—the homestead arts there came alive for us. There too we found that we had bodies and hands and feet as well as minds. We began to catch glimpses of the beauty in the concepts of life's oneness and wholeness. No aspect of human thought or activity remained untouched. Our understanding of our places in all fields deepened and came alive with promise.

A constant stream of student visitors flowed through the doors, mostly for seminars or short stays. In kitchen, garden, barn, and mill we soon became adept at both instructing and demonstrating. Whenever possible we let the students learn by doing, suffering silently within if one of them occasionally weeded all the pumpkin vines from the corn rows. Our journey had begun most encouragingly.

Here were intellectuals for whom telling and writing were not enough. They were doing things with their own two hands. Growing, grinding, baking, preserving, building, weaving—homesteading, they called it—they were actually controlling much of their own lives. They even printed some of Ralph's books themselves. Little did we realize then what a lifelong ecstasy awaited us through these seemingly chance contacts.

Not far from Suffern, in Spring Valley, was Three Fold Farm. One of the centers of the Anthroposophical Society, its program was based on the teachings of Rudolph Steiner in Germany. Its practitioners called their organic gardening and farming practices biodynamic. Their approach to the soil was natural and wholesome. Representatives occasionally came over to lecture, always most enthusiastically, on this "new" theory. It all made so much sense that its general principles were adopted in the school garden. We began teaching composting to our students.

The farther we went, the more we realized that we were only at the delightful beginning. We just had to know more about how to plant, tend, and harvest field-scale crops—the grains, hay, straw; how to manage farm-scale husbandry of animals and poultry; how to work with horses and farm machinery. In short, to our knowledge of the basic homesteading arts we felt we must add the ability to make our complete living from the soil. We had to do the job organically, of course, without synthetics, artificials, and manufactured chemicals, and in a manner that would enrich and enhance rather than deplete.

We learned after a time about Dr. Ehrenfried Pfeiffer who, before Rudolph Steiner's death, had worked with him for a number of years in Europe. Dr. Pfeiffer was an agriculturist and soil scientist who had come to this country a year or so earlier to establish an organic farm school. The very thing we now wanted most was there, waiting, in Kimberton, near Philadelphia, Pennsylvania.

We were always on the edge financially. The five dollars per week we received at the School of Living, in addition to board and room, had not greatly distended our purses, what with the arrival of the first baby and all that. But when the offer of fifty dollars a month for the three of us came through from Kimberton Farms School, out of which we had to pay all expenses but housing, we jumped at the chance. We were used to scrimping. When an infant cousin of our baby came to spend a year or so with us during illness in her family we knew we would have to pull our belts tighter. It was not easy to leave our Suffern home.

The one-thousand-acre farm project at Kimberton was run almost entirely by the ten to fifteen men and women students. It was tremendously hard work. Aside from the early tractors, modern farm labor-saving machines were just coming into use. We did much of our work with horses and by hand. We learned the cycles of the soil and its life. The hardships and the penury ran off our young shoulders.

Only when we felt we had learned what we needed to know to go on our own as full-fledged farmers did we decide to leave. Sadly enough, not long after our departure the farm school closed, never again to open. What a time today would be for such a place.

By spring 1946 we had finally located a beautiful farm to purchase in the rolling hills of the central Pennsylvania Appalachians. We had quite a time raising (by borrowing) the five thousand dollars needed to buy this one hundred acres, with house, barn, and innumerable sheds. In March we moved there—two children, two parents, Betty's elderly missionary father, a team of horses, our dog Lassie, and an old car.

The early Pennsylvania German settlers looked for black walnut trees as indicators of the limestone soil they sought. Here in Penns Creek the farm is full of them. We carried on the farm name given by the former owners, in whose family it had been held for a hundred years or more: Walnut Acres. Some of these trees must have been standing when William Penn owned all the land hereabouts.

Never was new-born babe more beautiful to a relieved mother than was Walnut Acres to us as we rattled proudly up the winding lane by the stream on that bright March moving day so long ago. Glory was everywhere. The tin roofs are rusted through in spots? Set buckets under the drips until we find time to patch the holes. The house and barn haven't been painted for twenty years, the windows are falling out? Ah, but the wood is sound—and just paste paper over the holes for now. The place has no plumbing, no bathroom, no telephone, no furnace—we must heat with a wood-burning stove? That's all right. Isn't it great to pioneer? We must pay off the mortgage with that one team of horses, plus an old plow and an old harrow—and live besides? Tut, tut—we've lived on nothing before; we wouldn't know how to live otherwise. Oh, the wonder of it all. We had a house and barn and outbuildings and a hundred acres. Did you hear? One hundred acres!

The years went by. Soon there were three children. Then two teams of horses. Then the horses left and a tractor came. We often had cows and sheep and chickens and guinea hens and ducks and geese and bees and cats and dogs. How soon fields and people and livestock all grew together into one family, each member knowing and accepting the other, albeit sometimes on sufferance. We became one mutually helpful unit. We knew where our priorities and our interdependencies lay. We respected each others' needs. Each of us tried to give more than we took. Such a husbanding, nonexploitive partnership has elements of the marvelous about it. We must have felt something of what countless country people, peasant folk, have felt in unbroken line since time immemorial.

All the time we were falling in love with our fields we were being made aware of increasing outside interest in their returns. People, first in ones and twos, then in ever-growing numbers, wrote to us, or came to see us at work. They wondered if they might purchase foodstuffs raised as we were raising them. The whole approach and procedure made such elemental sense to them that they were willing to pay the extra cost involved in creating and maintaining healthy, balanced soil, with its resultant balanced, tasteful produce.

A first postcard, then a first letter, then a visitor came from New York City, which was two hundred long, winding, rough miles away. I was sitting on the old barn roof, painting over the rust at last when that first auto appeared. I almost fell off.

We were one of the first to enter into this type of work, and so our name has become well known to many in this country, and even in some areas overseas. We did not plan this renown. We wanted originally just to get away from that crescent stretching from Boston to Richmond—to live simply and quietly, raising our family in a typical conservative rural society. But we soon found ourselves caught up in something much larger than we were, and we would have felt remiss not to give it the best we had.

Small decisions, coming one at a time, have a way of quietly adding up to something larger with the years. Looking back, we are amazed at the results of our having taken some of the turnings so long ago. Yet if in total these choices have led to something of universal significance, which we think they have, who will dare question either past steps or the mystery that opens up paths and directs one's feet into them?

Now you will find a fair-sized group of us, friends and neighbors, working together generally in amity and fellowship, producing in great variety and assortment foods fit to eat. By pooling talents of us all we have been able to build our program upon that often least-valued but most necessary element, common sense. We have learned both on the farm and in our enlarged farm kitchen and storehouse how to grow, harvest, store, grind, bake, roast, toast, can, freeze, package, and ship hundreds of food products, some of them unique.

The ultimate consumer has responded. We have had thousands of letters telling us what better, whole-food products have meant to individuals and families. We have been cheered by the occasional appearance in our midst of third-generation Walnut Acres-fed children: blooming, aware, alert, alive. Their bones are our stones. Their flesh is our labor. A tremendous testimonial; a burden of responsibility. To see the results of one's efforts in living beings, as we have been permitted to do, is one of the greatest of life's heartwarming privileges.

We who work here are not organized into a special type of community. We are individuals and families living apart from one another in the larger rural community of which we are members. The way in which we operate does not depend upon special or unusual living conditions or philosophical leanings. The same approach to our business affairs could work anywhere. But we have built into our work a specialness that we feel has interest and merit.

After working with us for two years, happily and satisfactorily, a person becomes eligible for membership in Walnut Acres, Incorporated. Five shares of stock are given to members each year for twenty years, to a mandated maximum of a hundred shares. The corporation owns the buildings and the land on which they are placed, the stock, inventory, formulas, mailing list, production methods, and so on.

The workers-stockholders thus own the business, and they share in its risks and its glories. The former seem often to outweigh the latter. And while the work goes on in its day-to-day fashion, at the heart of it all is a goodness and rightness and a fairness that seems to make for genuine happiness and satisfaction.

We have the opportunity to grow with our work, each person to his or her limits. It is our venture, our business, our productive unit. Together we help to feed thousands of persons with the best of foods. A purpose runs through it all and ties us and the universe into a tight, meaningful bundle. In this unit, both soil and people are the most important things on Earth. Finally, through the dedicated Walnut Acres Foundation, which both we and our customers help to support, we reach out soundly in uplifting ways to less fortunate persons both in our area and around the world. Life is all one piece. Through our entire program we try to see it whole.

Walnut Acres has meant much to us as a family—to Betty and me, our three daughters and their families. We feel we have been able to do so much more than we could ever have done alone as an individual homesteading family. Occasionally we look back, perhaps even wistfully, to the relative quiet and simplicity of our wood-burning, horse-drawn days. But we have really had the best of two worlds, and we would not want to have missed any part of either.

By being close to the center of the natural food and farming movement for many years, first helping hold it together, and later sharing its strong forward thrust, we've seen the future sweep into the present with a tremendous rush. Let's hope that the renewed awareness represented by this movement will be one of the saving elements so desperately needed now, to lead us moderns away from ourselves and our unbearable cocksureness into a quiet, thankful understanding and acceptance of life's limits and controls.

We've wondered sometimes about our growth. We would not

want to grow so large as to limit our ability to apply our idealism fully to all portions of our work. We do not think we have done so; we don't know how large we could grow and remain sound. We feel of late that we have come close to a size that seems ideal.

That all of us are genuinely involved as workers in every aspect of the work seems to give the farm more of solid base, making size itself less frightening. Yet over the years we have fought off growth for growth's sake. Each move, each chasing out of animals to make way for people, each replacing of hay with piled-up cartons of goods has wrenched our spirits. After careful weighing, the reasons for taking each step seemed ever more valid than those for remaining as we were. The way led where we had not expected it to; we tried to follow, but not without occasional doubts and fears.

We still think small. We live one day at a time, plant one plant at a time, harvest one crop at a time. Our fields are small. They average less than three acres each. Our village is small. We are perhaps four hundred residents now. Our church is small, our schools, our groups likewise. It may be that the richness of life resides in smallness. People, not machines, are at the center. Each man, each woman is a person, and not just a portion of the whole.

And so Walnut Acres is simply an extension of the diversified, family farm idea. All around us lie good farms, on some of which live frugal Amish families. Most farms in our area have from one hundred to two hundred acres. All are family farms. By careful planning, hard work, and simple living, the families carry on. Because of hills and stones and small fields and narrow contour strips, it is more costly to operate farms here than in Iowa. Average yields here may also be less because the topsoil is not as deep. If market prices are set so that the Iowa farmer can just make a decent living, our area farmers really suffer.

They hang on, even in lean years, just because they love it. The Iowa agribusiness farmers must now face investments of a million dollars or more. (How they must push the soil to get that back.) Land is high-priced enough here so that a young person without means can scarcely go to farming even now. Yet the solid virtues of the small farm remain. The time may yet come when we will see our country's greatest task as that of making land available to those who would use it wisely, on a small scale. We may learn to order things so that millions of families can once more make a substantial portion of a complete, satisfying living from the soil. Perhaps we can once again realize the virtues of simplicity and frugality.

As we see it, most people on this Earth gain strength from closeness to the soil and from thinking in small terms. Perhaps these are both basic needs, on which continuing life on this planet is predicated. Perhaps it has always been this way, and we are just relearning what those who went before already knew. In the midst of so many rapid changes that toss us about like leaves in the wind, it is good to know that Earth abides, and that small is beautiful.